MASTERING TELEVISION TECHNOLOGY

A CURE FOR THE COMMON VIDEO

COLEMAN CECIL SMITH, P.E.

Newman-Smith Publishing Company, Inc.

Richardson, Texas

Newman Smith Publishing Company, Inc.
2116 East Arapaho Rd., Suite 479
Richardson, Texas 75081 USA
(214) 231-6804

First Edition
Second Printing 1990

Library of Congress Catalog Card Number: 88-90786

ISBN 0-929549-02-3

Cover Design by Mark McGahan

Printed in the United States of America

TABLE OF CONTENTS

CHAPTER 2 THE BASICS OF AUDIO

CHAPTER 3 RECORDERS

CHAPTER 4 VIDEO SIGNAL CONCEPTS

CHAPTER 5 TELEVISION SYSTEMS

Mastering Television Technology

FOREWORD

Behind all the glitter and glamour of television production is a specialized technology that must be carefully adjusted to provide the highest possible quality of picture and sound. Just as paint, canvas, clay, and metals are media that are manipulated by an artist trying to precisely communicate his ideas; television technology is the medium that must be manipulated by television production personnel. *Mastering Television Technology: A Cure for the Common Video* dicusses applied video and audio technology for the real television production world.

Mastering Television Technology is for television directors, producers, and other production personnel; however, it is not a book intended to turn a director or producer into a technician. The intent is to introduce technical concepts so that production personnel can adjust equipment to squeeze everything out of the television medium and be able to effectively communicate with technicians when something goes "worng." Discussions employing mathematics and component-level electronics have been minimized. Operational concepts and logical development of block-level equipment discussions have been maximized.

Mastering Television Technology reflects the latest in technology and technological trends at the time of publication. It is a time of transition from tubed cameras to solid-state cameras, from conventional whole-signal recording techniques to component recording, and from analog to digital signal types. Much of the mystique and technical jibberish of the current and new technology is carefully explained to help equipment purchasers make intelligent technological decisions.

Many, many people have contributed to the development of this book, both directly and indirectly. Thousands of students in video technology seminars over the last fifteen years have pleaded for its development. The initial reviewers - Barbara Haley, Marvin Holland, and Rich Chambers worked long and hard to keep me on the right track as the book developed. Steve Cartwright, Jerry Hodges, Tom LeTourneau, Lon McQuillin, and Dick Reizner not only encouraged the development of this book, but also abetted the act by serving as final reviewers. Last, but not least, a big THANK YOU to Lynn (definitely my better half) in putting up with the hassles of book development for almost a decade!

CCS
Richardson, Texas
August, 1988

INTRODUCTION

Television technology is a science that has been applied to the arts. The drastic conceptual differences between using scientific equipment and generating artistically pleasing programming ,have created a dilemma in the television industry. How do basically unscientific artists adjust technologically sophisticated equipment for picture and sound that are of a consistently high quality?

These radically different concepts have made the distinction between "production" and "engineering" a long-standing division in the television industry. Two types of individuals have traditionally been involved in the process of creating a television product. One type of person made sure that the picture was framed properly, that the sound created the desired aural scene, and that the transitions from scene to scene evoked the desired feelings. A second type of person made sure that the equipment was adjusted and maintained to optimum technical performance.

The advent of small, economical television equipment has made some radical changes in the two-person technical/artistic concept throughout the television industry. Television production crews are getting smaller and smaller while experienced technicians and engineers are becoming more scarce. This dynamic combination dictates that each member of a crew; whether specializing in production, engineering, or management; must understand the entirety of what is required to create quality television programming.

Mastering Television Technology attempts to explain what is going on inside television equipment in a logical, easy-to-understand fashion for producers, engineers, and managers. Television production personnel must understand the effects of all the adjustments that they make to equipment. The infamous little green screwdriver needs some intelligent guidance to minimize the number of times that it introduces new problems. An appreciation of

1

the technology involved provides an insight into what is going on inside television equipment so that adjustments are not made on a hit-or-miss basis. Understanding television technology also requires exposure to technical terms, allowing production personnel with technical problems to effectively communicate with technicians.

Television technical personnel need an occasional review of technology to keep their skills sharp. Even with long experience in technical television, many technicians need to fill a few voids in their experience.

Managers responsible for television personnel and technology need to understand how equipment and changes in technology enhances operations under their charge. Improvement of picture and sound quality, increased throughput, reduced personnel costs, and reduced operational costs are a few of the benefits that must be weighed against the significant capital costs of equipment for television activities.

Mastering Television Technology may be read cover-to-cover as a comprehensive overview of television technology; as a reference, using the comprehensive Table of Contents and Index to pinpoint specific topics; or as a textbook in support of television technology curricula.

Each of the five chapters explores a major area of television technology which must be understood to create the best picture and sound with the least effort. "The Basics of Video" discusses what it takes to generate and display a color video signal for a "live" production. The concepts of live production must be understood before beginning to understand recording, switching, processing, or distributing the video signal.

"The Basics of Audio" covers general concepts of live audio signal generation and presentation. Audio technology may appear curious in a book that professes to "cure the common video," but the discussions introduce technical concepts that are common to both audio and video circuits. These concepts are usually encountered in the real world when trying to match signal parameters while interconnecting audio equipment. Video systems are easier to interconnect because of standardization of the same signal parameters.

"Recorders" discusses the process of video and audio recording. The various techniques used to maintain a quality color picture and sound through the distortion-prone recording processes are discussed from a conceptual viewpoint.

"Video Signal Concepts" closely examines the composition, measurement, and correction of the video signal. It is the video signal that a monitor displays as a picture. If the video signal is not correct, there is no way in which the picture can be correct.

"Video Systems," discusses the various pieces of supporting equipment that must be properly interconnected to create a television program. Discussions about video switchers, character and graphics generators, and film chains cover the operational concepts of equipment frequently seen in video systems. System timing and testing discussions explore the way in which the video signal must be treated to maintain signal stability and quality as it progresses through a system.

The only prior knowledge required to understand the contents presented in *Mastering Television Technology* is obtained with experience in the television production environment. Readers who realize what they need to know will have the easiest time wading through the diverse details of television technology. Until you try to correct recorded color balance errors, there is no motivation to understand red, green, and blue gain adjustments in a camera. Until you hear those terrible scratchy esses in your audio, trying to understand level, impedance, and balance matching of audio equipment isn't a topic of overwhelming concern.

Although it helps to think like an electron, prior in-depth knowledge of electronics, physics, and mathematics is not required to understand most of the discussions. Many people shudder with fear at the thought of these subjects. Among the primary goals of *Mastering Television Technology* is alleviation of this fear of television technology by filling the void of competent and understandable information and dispelling some of the myths about television technology.

Many of the fears of television technology stem from the increasing rate at which technological advances have been arriving. The advances make the equipment easier to operate, lower in cost, and capable of producing sound and picture that is higher in quality. On the other side of the coin, these same technological advances have made many television operations much more complex to comprehend.

The motivation behind many of these technological advances has not been entrenched in direct user benefits, but in acceptance or forecasted acceptance of the advance in the professional equipment market. There have been several video tape formats that have withered simply because nobody bought them. Some manufacturers survived only because they offered revolving credit plans to their customers. This marketing drive is reflected in many of the "technical" specification sheets with confusing, inconsistent, and frequently misleading information. Hopefully, *Mastering Television Technology* will clear away some of this smoke.

CHAPTER 1

THE
BASICS
OF
VIDEO

LIGHT AND LENSES

The television system converts the light information from a scene into an electrical "video signal" for transmission of the scene's visual image to a remote location or for storage on magnetic tape for later use. Before looking at the processes that take place in the conversion from light to electricity, examination of some of the basic characteristics of lenses can help understand the nature of light and images.

There are two types of light with which the television system must deal, "incident light" that falls on the objects in a scene and "reflected light" that bounces off the objects. The reflected light is the portion of the incident light that allows perception of visual details of the object.

The intensity of light that illuminates an object is usually measured in English units of "foot-candles" (f-c) or in Metric units of "lux" (1 f-c = 10.76 lux). The intensity of light that is reflected off an object determines the brightness of the object and is frequently measured in "lamberts" where 1 lambert = 0.318 candles-per-square-centimeter. Another unit called a "lumen" denotes the intensity of "luminous flux" where 1 f-c = 1 lumen incident per square foot. Additional information about light measurements may be found in most basic textbooks about physics.

7

When light falls on an object, it is reflected off each tiny area of the object's surface in many different directions. An eye collects some of the reflected light and perceives it as a visual scene. As shown in Figure 1.1, if the light reflected from these objects is not carefully controlled, the eye sees the result of combining the brightness of all the reflections off all the surface points on the object. The eye can not see any sharply focused images, but senses the *averaged* total of incident and reflected light. If a sharply focused image of the object is desired, the light must be carefully controlled to ensure that the reflected light from only one point on its surface reaches a given point on the light sensitive surfaces within the eye. The eye uses the light-bending properties of a transparent lens to accomplish this control.

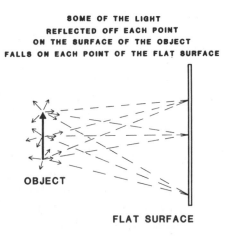

Figure 1.1
Uncontrolled Light Reflected Off an Object

Whenever light passes through a single material; be it air, water, glass, or whatever; it travels at a speed that is dependent on the characteristics of the material. When the light passes from one material to another, the speed of the light changes. A close examination of Figure 1.2 reveals that when a beam of light strikes a material transition at an angle other than perpendicular, the portion of the light waves within the beam that have already passed through the transition are bent relative to the remainder of the light waves that have yet to reach the transition. This bending or "refraction" alters the original path of the light beam.

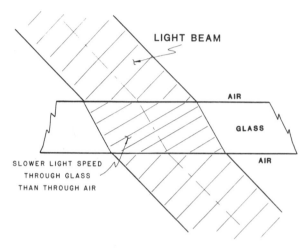

Figure 1.2
Light Refracts When Passing Between Materials

LENS FOCAL LENGTH

If a piece of transparent material is formed into a special shape, the refraction processes can be used to control the direction of the light. For example, a piece of transparent glass formed into a "convex lens" as shown in Figure 1.3 changes the path taken by many of the reflected light waves from a point on the surface of an object and focuses them at a point some distance behind the lens. If the thickness of the lens is increased, the distance between the object and the focused image decreases. The distance behind the lens where the image is focused is called the "focal length" of the lens. If a flat surface is placed at the focal length distance behind the lens, a sharply focused image of all of the visual details of the object in the scene is visible.

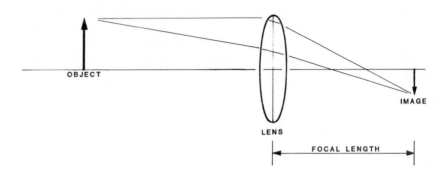

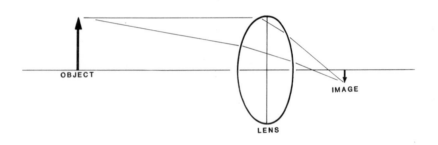

Figure 1.3
Lens Focal Lengths

If a constant size of detected image area is maintained and use several lenses in a cylindrical assembly, the focal length can be used to determine the distance between the lens and the object within the "angle of view" limitations imposed by the lens. As shown in Figure 1.4, a short focal length lens has a wide angle of view, allowing a stationary object to occupy less of the framed picture than a lens with a narrower angle of view.

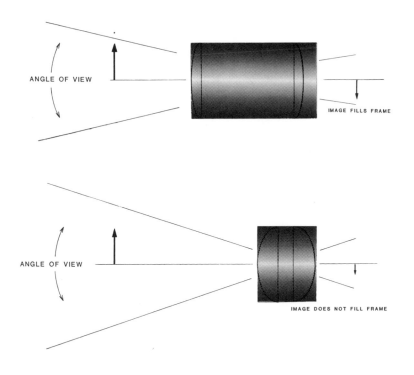

Figure 1.4
Lens Focal Length Determines Angle of View

In describing the visual characteristics of lenses,
the focal length is usually specified in millimeters
(mm). Thus a 50 mm lens has a wider field of view
than a 100mm lens if the size of the focused image is
the same. But if the size of the focused image
changes, the relationship is not valid. For example, a
25mm "normal" lens that focuses a 1-inch in diameter
image has a 23-degree angle of view. If the image size
were changed to 2/3 inch, the focal length for the
same angle of view becomes 23mm. Any focal length
longer than the "normal" for a given focused image
size is considered a "telephoto" lens. Any shorter focal
length is considered to view a "wide angle" of the
scene.

11

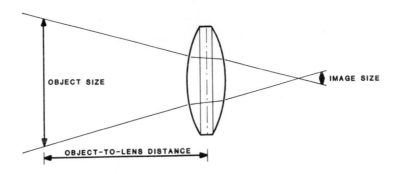

Figure 1.5
Determining the Focal Length of a Lens

Referring to Figure 1.5, the focal length is related to the width (or height) of the focused image and the corresponding size of the viewed field by the following formula:

$$\frac{\text{Focal Length}}{\text{Image Size}} = \frac{\text{Object-to-Lens Distance}}{\text{Field Size}}$$

Thus, to view a 15 foot field width at a 30 foot distance from the lens and focus that into a 0.8 inch image width requires a focal length of:

$$\frac{\text{Focal Length}}{0.8 \text{ inch}} = \frac{30 \text{ foot X 12 inches-per-foot}}{15 \text{ foot X 12 inches-per-foot}}$$

$$\text{Focal Length} = 2 \text{ X } 0.8 \text{ inch}$$

$$= 1.6 \text{ inch X } 25.4 \text{ mm-per-inch}$$

$$= 40.64 \text{ mm}$$

THE TELEVISION FRAME

In television, the image passed through the system must conform to the shape of a rectangle with proportioned side lengths specified by the "aspect ratio." In most television systems, the aspect ratio specifies that the picture must have three units of height for every four units of width. Thus, an image that is three feet high has a width of four feet. Special anamorphic or "wide screen" television systems may use other aspect ratios. Since most glass lenses used in television are round, some of the round focused image that comes through the lens must be discarded. To optimize the quality of the image, the amount of image that is discarded must be minimized. As shown in Figure 1.6, this means that the rectangular television frame must be as large as possible within the round focused image. This can be achieved if the diagonal measure (the longest distance between any two points within the rectangle) is equal to the diameter of the image. To find the width of the frame under these circumstances, the rectangle can be divided into two right triangles and the Pythagorean theorem can be used where:

$$A^2 + B^2 = \text{Hypotenuse of Triangle}^2$$

$$(3/4 \text{ Width})^2 + \text{Width}^2 = \text{Diagonal Frame Size}^2$$

$$\text{Width}^2 = 16/25 \text{ X Diagonal Frame Size}^2$$

$$\text{Width} = 4/5 \text{ X Diagonal Frame Size}$$

Since the diagonal frame size equals the focused image size under optimum conditions,

$$\text{Frame Width} = 4/5 \text{ X Focused Image Diameter}$$

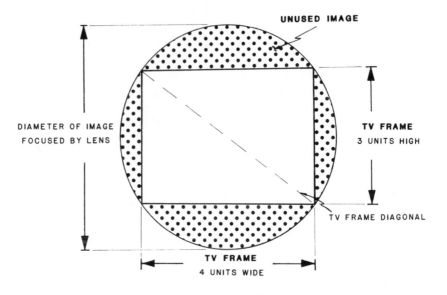

Figure 1.6
TV Frame Inside a Focused Image

Courtesy Barco Industries

Figure 1.7
Monitor Display Using an HDTV Aspect Ratio of 5:3

For a 1-inch image diameter,

> Frame Width = 4/5 X 1 inch
>
> = 0.80 inch

For a ½-inch image diameter,

> Frame Width = 4/5 X ½ inch
>
> = 0.40 inch

As light passes through a single glass lens, the various light frequencies (that are perceived by the eyes as various colors) is slowed down by varying amounts. Remembering that speed changes when light goes through a material transition results in bending the path of the light, these variances in speed of the various colors shows up in the focused image as color fringes around the objects in the focused image. To correct these "chromatic aberrations," lens manufacturers use computer-aided design and manufacturing techniques to fill a cylindrical lens assembly with several glass lens "elements" for optimum image sharpness and quality.

To focus an object located a specific distance from the lens and maintain color correction, the lens assembly contains a lens carrier similar to the zoom lens shown in Figure 1.8 that can be moved back-and-forth by a screw-thread drive system. The position of the lens carrier is controlled by an adjustable "focus ring" on the outside of the lens assembly. Since this adjustment also changes the size of the focused image, the specified focal length of the lens is reached when the lens is adjusted to focus at infinity.

Zoom lenses, which can continuously vary the focal length through a designed range, frequently have eighteen to twenty individual glass lens elements and several independently moving carriers to maintain color correction and sharp focus of the image. Besides the maximum and minimum focal lengths, zoom lenses are also specified by the ratio of maximum focal length to minimum focal length. Thus a zoom lens with a maximum focal length of 100mm and a minimum focal length of 10mm has a 100/10 = 10:1 or 10X (ten times) zoom ratio.

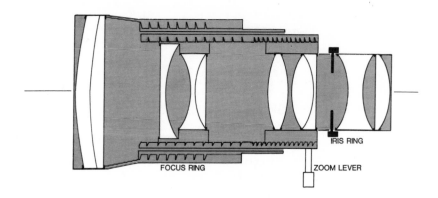

Figure 1.8
Typical Zoom Lens Cross Section

Figure 1.9
Typical Zoom Lens

LENS IRIS

Because television equipment must be operated within strict limits of light intensity, the amount of light passing through any lens must be carefully controlled. Usually the amount of light is controlled by placing an "iris" (a diaphragm with variable diameter opening) in the lens assembly. This diaphragm is placed at a point in the lens assembly where the paths of the various reflected light waves are traveling separated from each other. At this point, the iris obstructs some of the light paths and passes the remaining light. Because some of the focused light reflected off the objects in the scene still passes through, the iris is usually out of focus but the cumulative amount of reflected light illuminating each point on the focus plane is controlled. In some lens adjustment conditions, particularly in low light and wide viewing angles, the iris may produce a darkening or "vignetting" at the corners of the focused image.

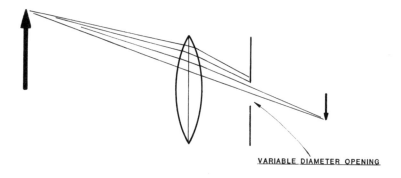

VARIABLE DIAMETER OPENING

Figure 1.10
Typical Lens Iris

Adjustment of the iris is provided on the outside of the lens assembly by an "iris ring" calibrated in "f-stops." A change in the f-stop value is inversely proportional to the amount of light passing through the lens (higher f-stop = lower light transfer). Mathematically, the f-stop value depends on the iris and the focal length as follows:

17

$$\text{Iris Opening Diameter} = \frac{\text{Focal Length}}{\text{f-Stop Value}}$$

Thus, on a 25mm lens set to f8, the iris has an opening of

$$\text{Iris Opening Diameter} = \frac{25\text{mm}}{8} = 3.125\text{mm}$$

$$\text{Iris Opening Area} = \pi(\frac{\text{Diameter}}{2})^2 = 7.67\text{mm}^2$$

If the f-stop is changed by one stop to f11 (the next value on the lens), the iris has an opening of

$$\text{Iris Opening Diameter} = \frac{25\text{mm}}{11} = 2.273\text{mm}$$

$$\text{Iris Opening Area} = \pi(\frac{\text{Diameter}}{2})^2 = 4.06\text{mm}^2$$

When changing between two stops marked on a lens, the area of the iris opening is changed by a factor of two (multiplied by two or divided by two). If the iris area is changed by a factor of two, the amount of light passage through the lens changes by a factor of two.

If the amount of light that falls on an object were to double from 100 f-c to 200 f-c, the lens may be "stopped down" from f8 to f11 (halving the amount of light passing through the lens) to adjust for the same amount of light in the focused image.

The iris may be automatically adjusted by a motor that is driven to maintain a constant *average* voltage of the output video signal. This "auto iris" capability allows easy and consistent operation under widely varying operating conditions. But auto iris is not able to pick out and adjust for picture details that are of substantially different brightness than the average brightness of the entire scene. For example, if a person is facing into a room in front of an open window and we wish to see details of his face, the average amount of light is bright because the light from the window in thebackground forces the automatic iris to adjust to the point where the head is just a silhouette. To retrieve the details in the face, the iris must be manually adjusted to overexpose the background.

DEPTH OF FIELD

Focus, focal length, and iris affect the *range* of distances from the lens that an object is in acceptable focus. This "depth of field," as shown in Figure 1.11, varies with focus distance, focal length, and f-stop in the following relationships:

DEPTH OF FIELD IS DIRECTLY PROPORTIONAL TO FOCUS DISTANCE.

As the object gets closer to the lens, the range of distance that the object is in focus gets smaller.

DEPTH OF FIELD IS INVERSELY PROPORTIONAL TO FOCAL LENGTH.

As the focal length of a lens is increased, the range of distance that the object is in focus decreases.

DEPTH OF FIELD IS DIRECTLY PROPORTIONAL TO f-STOP.

As the f-stop number is increased (less light allowed to pass through the iris), the range of distance that the object is in focus increases.

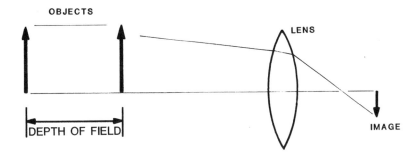

Figure 1.11
Depth of Field

In television production, it is not unusual to use these lens relationships to creative advantage. For example, if a shallow depth of field is desired to visually separate objects while shooting in sunlight, a colorless neutral density (ND) filter can be placed in the path of the light to reduce the amount of light passing through the lens (without introducing any color changes), therefore requiring a larger lens opening and reducing the depth of field. Such techniques are frequently used when a viewer's eyes are to be focused on a particular foreground object separated from an unimportant background.

If the depth of field is too shallow to allow two objects at different distances to be in the sharpest focus, the illusion of distance separation may be enhanced if the lens is focused at a point behind the closest object that is approximately 1/3 the distance to the farthest object. This produces a sharper close object which the viewer's brain perceives as being closer to the lens.

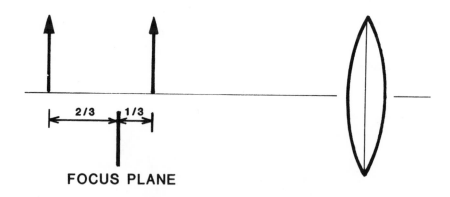

Figure 1.12
Focusing Between Two Objects

GENERATING THE VIDEO SIGNAL

As shown in Figure 1.13, if the lens is used to focus its image on a flat surface coated with a material that changes its electrical characteristics with varying light, the light information can be readily converted into an electrical video signal. There are two basic methods of generating a video signal by referencing the focused image: solid-state devices and tubes. Solid-state pick-up devices use a mosaic of individual light-sensitive elements generate charges that vary with incident light.

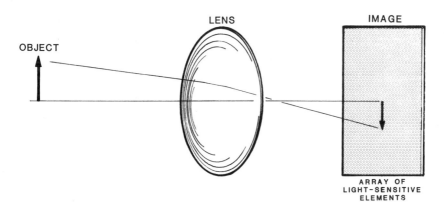

Figure 1.13
Focusing the Image on a Light-Sensitive Matrix

Figure 1.14 shows how tube-type television pick-up devices use the focused image on a coating of light-sensitive target material that varies resistance to electricity trying to flow through it. A constant flow of electricity, in the form of an electron beam, is generated within the tube to pass through the light-sensitive target material. As the electricity flows through the material, it encounters varying electrical resistance in response to the amount of light. Bright light produces a "low" target resistance and allows a large amount of the electrical flow to get through.

Dim light produces "high" target resistance that allows only a small amount of the electrical flow to pass. Using this technique, the output strength of the electrical current is altered to correspond to the intensity of light that falls on the target material. To get details about all the available brightness information, the focused image is "scanned." (A detailed discussion about scanning appears later.) To allow optimum performance, the electron beam and target are enclosed in a glass envelope from which most air and other contaminants have been removed.

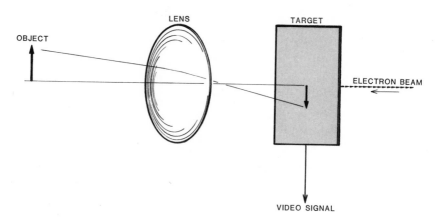

Figure 1.14
Target Material Varies Electrical Flow

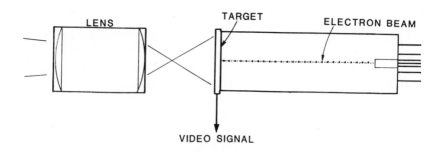

Figure 1.15
Electrical Flow Through a Pick-up Tube

CHARGE COUPLED DEVICES

Charge coupled devices (CCD) are frequently used to convert the image into a usable video signal. CCDs were developed as a replacement for pick-up tubes to drastically reduce power consumption, weight, and size. As shown in Figure 1.16, the size of cameras that use solid-state imaging are smaller than conventional pick-up tube cameras.

Courtesy Panasonic Industrial Company

Courtesy of Sony Corporation

Courtesy BTS Corporation

Figure 1.16
Typical CCD Cameras

FRAME TRANSFER CCD

CCDs use thousands of light-sensitive elements in a matrix to detect the light intensity at various points in the focused image. The brightness of a given image detail determines the amount of electrical charge generated by an imaging element. These charges remain in the "imaging" elements where they are generated until control signals command the video charges to transfer to an adjacent "transfer" elements. The transfer elements used the same control signal to command any existing charge to vacate and make room for the new charge.

This technique of charge transfer is frequently called "bucket brigade" because the technique is similar to the way fire fighters used to operate. Buckets of water were transferred from one brigade member to the next to fight the fire. A CCD transfers varying charges (the amount of water in a given bucket) from one matrix element to the next (one brigade member to the next) to create the video signal.

This transfer of charges within the CCD effectively creates a "scanning" of the focused image. During charge transfer, the CCD must be shaded from light or insensitive to light because the matrix elements used to transfer charges sometimes double as imaging elements.

In a "frame transfer" CCD, as shown in Figure 1.17, the charges that represent the visual information are rapidly transferred from the imaging matrix into a second matrix for temporary storage. The charges in the temporary storage matrix are then "slowly" shifted out of that matrix into an adjacent output transfer matrix before being output from the device.

The transfer from the storage matrix to the output matrix happens at a rate corresponding to the vertical scanning rate of the television system. One set of charges, corresponding to one horizontal scan of video information, are transferred to the output matrix at one time. All of the charges in the storage matrix are transferred to the output matrix in 1/60 second to effect vertical scanning.

It may help to visualize the charge transfer process by considering the spring-loaded salad plate holder found at some restaurant salad bars. If a plate is removed, the next plate on the stack springs up to take the place of the removed plate. If the charges of one horizontal scan of video (one salad plate) are removed, the charges of the next horizontal scan of video (next salad plate) is transferred (by the spring in the holder) as needed.

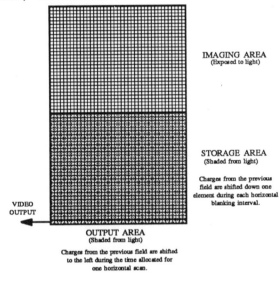

IMAGING AREA
(Exposed to light)

STORAGE AREA
(Shaded from light)

Charges from the previous
field are shifted down one
element during each horizontal
blanking interval.

VIDEO
OUTPUT

OUTPUT AREA
(Shaded from light)

Charges from the previous field are shifted
to the left during the time allocated for
one horizontal scan.

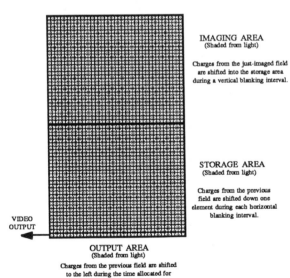

IMAGING AREA
(Shaded from light)

Charges from the just-imaged field
are shifted into the storage area
during a vertical blanking interval.

STORAGE AREA
(Shaded from light)

Charges from the previous
field are shifted down one
element during each horizontal
blanking interval.

VIDEO
OUTPUT

OUTPUT AREA
(Shaded from light)

Charges from the previous field are shifted
to the left during the time allocated for
one horizontal scan.

Figure 1.17
Frame Transfer CCD Operation

The frame transfer CCD cycle basically goes as follows:

1) The light sensitive matrix is exposed to the image all at one time.

2) During vertical blanking (the time normally used to retrace the scan from bottom-to-top), the visual matrix is protected from light by a mechanical or electronic shutter and the information is rapidly shifted to an adjacent temporary storage matrix.

3) After vertical blanking, the visual matrix is again exposed to the light information of the focused scene while the information that is stored in the temporary storage matrix is "slowly" read one scan at a time (to create an effective $262\frac{1}{2}$ horizontal paths scanned 60 times per second).

INTERLINE TRANSFER CCD

An "interline transfer" CCD, as shown in Figure 1.18 works using the bucket brigade concept but light-insensitive transfer elements are placed in the image matrix adjacent to the image elements. The charges representing the image brightness are shifted to the adjacent transfer element and down into an output register. The adjacent transfer elements improve response to rapidly changing scene brightness but reduce the ability of the CCD to sense image detail on the horizontal picture axis.

This charge-transferring "scanning" technique used by CCDs allows the operator to vary the exposure time of the light-sensitive area without altering television scan rates. This variation of exposure time allows alteration of the "shutter speed" in the same way as a photographic camera. In this manner, each field of video signal may be adjusted for image sharpness *or* for low-light operation, as required by the application. Sharp images in each field allows use of television in motion analysis and "slow motion" applications previously requiring the use of photographic motion picture techniques.

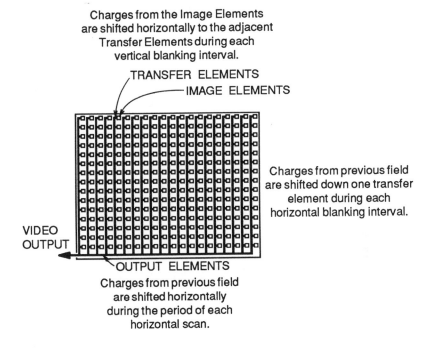

Charges from the Image Elements
are shifted horizontally to the adjacent
Transfer Elements during each
vertical blanking interval.

TRANSFER ELEMENTS
IMAGE ELEMENTS

Charges from previous field
are shifted down one transfer
element during each
horizontal blanking interval.

VIDEO
OUTPUT

OUTPUT ELEMENTS
Charges from previous field
are shifted horizontally
during the period of each
horizontal scan.

Figure 1.18
Interline Transfer CCD Operation

The ability of CCDs to efficiently handle large contrast ratios encourages application where pick-up devices may be subjected to dangerous light levels. The CCD endures much more light without permanent damage than a pick-up tube. The light response, however, may be marred in interline CCD chips where a white line extends downward in the picture from a super-bright picture detail (like a light fixture). This is, in essence, the result of trying to transfer too much charge through a vertical transfer register and allowing it to "spill" into areas where it should not go. It is easy to visualize the charge as being similar to water being contained behind a dam. If the dam (matrix element) is not high enough to contain the water (the video signal charge), the water overflows downstream.

Even though CCDs use new technology to "scan" the image, the same lens and optical requirements associated with pick-up tubes apply equally to CCDs.

PICK-UP TUBES

As was discussed at the beginning of this section, pick-up tubes operate by sending a constantly flowing stream of electrons through a target material that offers varying resistance to electron flow. The electron beam in a pick-up tube has a very small diameter. In fact, the optimum cross-sectional size of the electron beam is one electron (about 0.00000000008 inches). This optimum size focuses the electron beam to examine the smallest possible area of the image focused on the target to maximize the tube's ability to resolve details of the focused image.

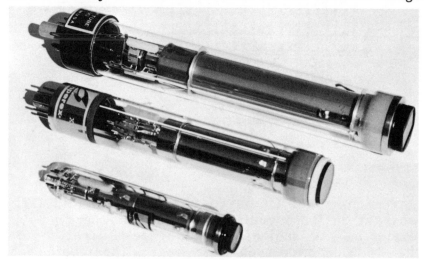

Figure 1.19
Typical Pick-up Tubes

Because of the small size of the electron beam and the larger size of the target material (usually ½-inch to 1-inch in diameter), all the details of a scene focused on a target cannot be examined at once. As shown in Figure 1.20, this problem is overcome by making the electron beam "scan" the scene left-to-right (horizontal) and top-to-bottom (vertical) to examine each picture detail, one after another. (This scanning convention is valid only when viewing the correctly oriented scene on a picture monitor. Either or both directions may be reversed within a camera to correct anomalies created by lenses, mirrors, etc.)

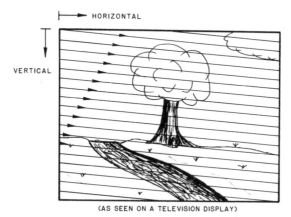

HORIZONTAL

VERTICAL

(AS SEEN ON A TELEVISION DISPLAY)

Figure 1.20
Scanning a Focused Image

As the electron beam scans the image focused on the target, the various brightness levels vary the electrical resistance as though the target material were made up of individual resistors, as shown in Figure 1.21. The capacitors drawn in parallel with the resistive elements in Figure 1.21 are a physical property of the target material that stores some of the electricity as it passes through and slows the response of the target to rapid variations in light.

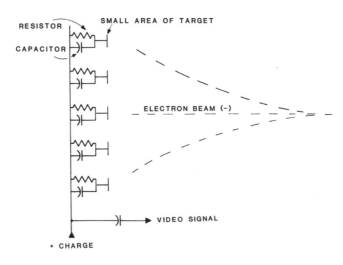

RESISTOR

SMALL AREA OF TARGET

CAPACITOR

ELECTRON BEAM (-)

VIDEO SIGNAL

+ CHARGE

Figure 1.21
Schematic Diagram of Pick-up Tube Operation

29

After completing each left-to-right horizontal scan across the target, the electron beam is momentarily turned off or "blanked" during its right-to-left return or "retrace" before it starts the next horizontal scan; likewise, the electron beam is blanked during the retrace from the bottom to the top of the target. Horizontal and vertical blanking eliminates picture interference and bright dots in the picture caused by what normally are overlapping scans.

Note that the "horizontal" scan is actually higher on the left side of the picture than on the right side of the picture. This is a result of the vertical scanning continuously pulling the scan downward while the many left-to-right scans are completed.

Figure 1.22 shows a graph of video signal voltage resulting from a typical horizontal scan. In this graph, the voltage at any instant is equivalent to the brightness of the picture details as they are encountered during the scan. The graph displays the voltage from one horizontal scan, left-to-right with the distance on the graph representing the left-to-right distance in the picture.

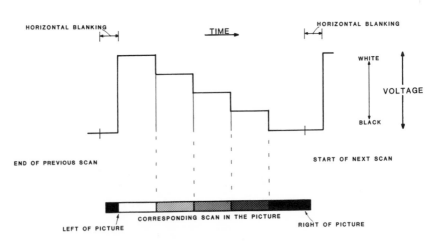

Figure 1.22
Brightness (Voltage) Graph from a Horizontal Scan

FIELDS AND FRAMES

Although some systems use a "progressive" technique to scan the entire scene in one vertical scan, normal television production equipment *vertically* scans the entire scene twice to generate the best possible picture quality. Doubling the vertical scan rate allows sandwiching of horizontal scan paths with the oldest picture information between paths with the newest picture information to reduce flicker, reduce any smear of a moving image, and allow a much faster response to drastic light changes. As shown in Figure 1.23, each vertical scan (of half of the picture information) is called a "field." One field contains every other scan path of the complete picture, or "frame." The other field contains only the remaining scan paths in the same frame. The two fields "interlace" to form a frame; thus there is a "2:1 interlaced" relationship of fields to frames in the television system.

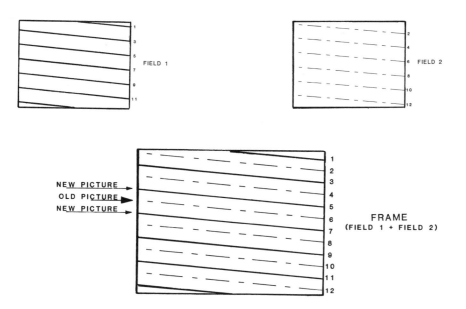

Figure 1.23
Two Fields Interlace to Form One Frame

SCANNING STANDARDS

In North America, the System M standard of The International Radio Consultative Committee (CCIR) of the International Telecommunications Union has been adopted, where a black-and-white system has:

525 horizontal scan paths per frame (complete picture)

$262\frac{1}{2}$ horizontal scan paths per field

30 frames (complete pictures) per second

60 fields ($\frac{1}{2}$ pictures) per second

Scanning system standards should not be confused with color encoding standards (like the NTSC standard discussed later), because there are many encoding and scanning system combinations adopted on a country-by-country basis. Program material produced in North America using System M standards may need to undergo "standards conversion" to be used in other countries. (CCIR System M should not be confused with the M or M-II video tape formats.)

ELECTRON BEAM SCANNING

Basic electrical responses of electrons are used to make the beam rapidly scan the target material. Electrons are attracted or repelled by *electric* fields which can be created by electrically charged metal plates or by coils of wire. If the charged plates or electromagnets are placed on the horizontal and vertical axes of a pick-up tube the path of the beam of electrons from the cathode to the target can be deflected. The angle of deflection of the electron beam as it travels through the pick-up tube can be varied by varying the strength of the electric field through which the electron beam passes. If the strength of the electric field is varied (by varying the voltage applied to the coils or electromagnets), the target can be scanned by the electron beam.

The charged plate technique shown in Figure 1.24, called "electrostatic deflection," places one set of metal plates above and below the pick-up tube for vertical deflection of the electron beam. Another set of plates to the left and right of the tube provides horizontal deflection.

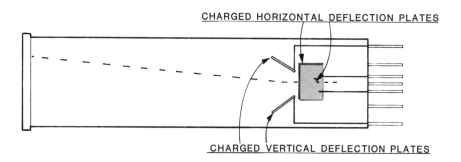

CHARGED HORIZONTAL DEFLECTION PLATES

CHARGED VERTICAL DEFLECTION PLATES

Figure 1.24
Electrostatic Deflection

As shown in Figure 1.25, when the "electromagnetic deflection" technique is used, the pair of electromagnets left and right of the tube create an electric field that deflects the electron beam vertically (60 times per second in North America). The pair of electromagnets above and below the tube create an electric field that deflects the electron beam horizontally 525 X 30 = 15,750 times per second so that each horizontal scan takes 1/15,750 = .0000635 seconds = 63.5 microseconds. The assembly that contains the electromagnets surrounding the pick-up tube is called the "deflection yoke assembly." Figure 1.26 shows how the constantly changing voltage used to create the vertical electric field in the deflection yoke assembly relocates the vertical position of the electron beam downward until the bottom of the scan is reached. Simultaneously, the horizontal electric field constantly relocates the horizontal position of the electron beam. Any physical relationship between the vertical scan and the horizontal scan is dictated by the applied deflection voltages.

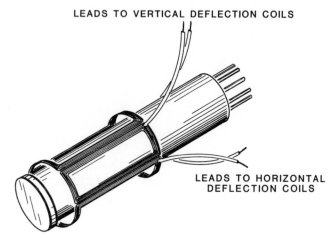

LEADS TO VERTICAL DEFLECTION COILS

LEADS TO HORIZONTAL
DEFLECTION COILS

Figure 1.25
Electromagnetic Deflection

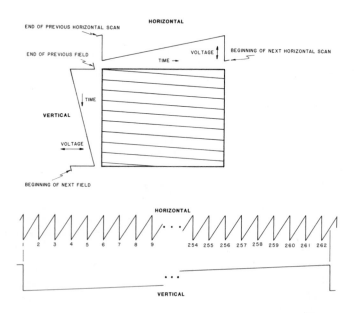

Figure 1.26
Signals Applied to Deflection Coils or Plates

TYPES OF PICK-UP TUBES

Historically, pick-up tubes have gravitated toward smaller and smaller sizes. Early tubes were up to seven inches in diameter, but as target materials with better picture detail resolving power were developed, tubes have now been reduced to as small as one-half inch in diameter. The most commonly used pick-up tube sizes include diameters of 1/2-inch (14mm), 2/3-inch (18mm), 1-inch (25.4mm), and 1.2-inch (30mm). *In general*, a reduction in pick-up tube size (while keeping the same target material) creates a corresponding reduction in detail resolution or light sensitivity. This is caused by the size relationship between aberrations in the target material and the scanned image.

Pick-up tubes currently in vogue include the vidicon, Plumbicon®, and Saticon®. Historically the vidicon is the grandfather of the small-format pick-up tubes. From that technology, other small-format tubes were developed.

The Plumbicon® was developed by N.V. Philips to provide an increased sensitivity to light (when compared to a vidicon) and stability of dark picture areas. As shown in Figure 1.27, the antimony trisulphide target material of the vidicons was replaced with layers of lead oxide (PbO), tin oxide, and "doped" lead oxide to form a "PIN" semiconductor.

The Saticon® was later developed by NHK, Japanese Broadcasting Corporation, to enhance the low-light characteristics of the vidicon while increasing picture detail resolving power. The target material in a Saticon® is composed of layers of selenium, arsenic, and tellurium.

PICK-UP DEVICE MEASUREMENTS

SENSITIVITY

Depending on the construction and selection of target material, pick-up devices produce pictures of varying qualities. Light sensitivity of a pick-up device is of great concern because the greater the sensitivity, the lower the noise (or "snow") in the finished picture. The sensitivity of a camera, which is highly dependent on the pick-up devices, is described as the minimum amount of light required to make a

35

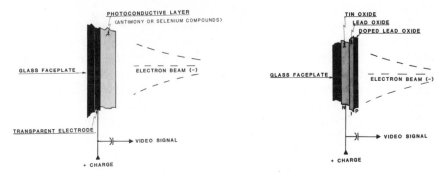

a) Vidicon Target   b) Plumbicon® Target

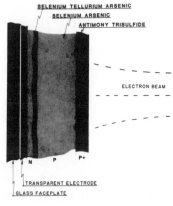

c) Saticon® Target

Figure 1.27
Pick-up Tube Target Construction

usable picture under given lens, noise, and signal level conditions. Normally a camera's required minimum level of scene illumination is measured in foot-candles (f-c) or lux. Figure 1.28 shows the specifications of a typical type of pick-up tube, as the engineer receives them. The sensitivity of the tube is still described there, just expressed in different terms.

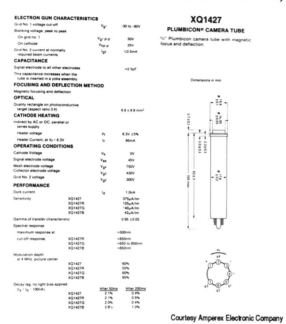

Courtesy Amperex Electronic Company

Figure 1.28
Typical Pick-up Tube Specification Sheet

RESOLUTION

Another factor of pick-up device selection is picture detail resolving power, or "resolution," which determines the sharpness of the output picture. The design of the pick-up device, a camera's pick-up device control settings, target material, and manufacturing tolerances found in a given pick-up device contribute to variations in resolution. As shown in Figure 1.29, horizontal resolution is frequently measured as the number of individual black/white *vertical* lines that can be discerned in the picture. This optical measurement is usually performed using a wedge-shaped pattern of converging lines similar to that shown in Figure 1.30. The point where individual lines in the wedge cannot be discerned indicates the limiting resolution of the system with the scale adjacent to the wedge indicating the number of lines of that size that could be placed on the screen. (A picture monitor can also contribute to loss of horizontal resolution, so a high resolution monitor or an oscilloscope is normally used to measure the resolution of a pick-up device.)

37

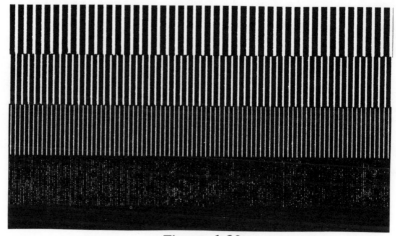

Figure 1.29
Number of Individually Viewable Lines
Expresses   Horizontal Resolution

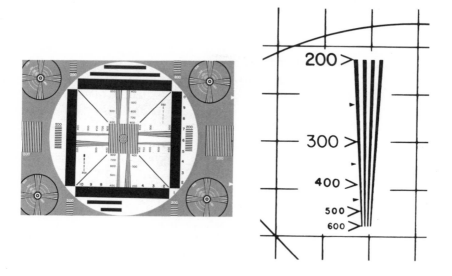

Courtesy Electronic Industries Association

Figure 1.30
Resolution Measurement Chart

Another way to describe pick-up device resolution is by expressing the "depth of modulation" of the pick-up device. Figure 1.28 shows that when using a type XQ1427 Plumbicon® manufactured by Amperex, 60% of the maximum black-to-white voltage amplitude remains when the voltage varies at a rate of four million cycles-per-second.

Figure 1.31 shows that the maximum number of transitions between white-black that occur on an given horizontal scan is determined by the upper frequency limit of the voltage response of the television circuits. Thus high-frequency response limit of the pick-up device and electronics defines the available picture detail resolving power.

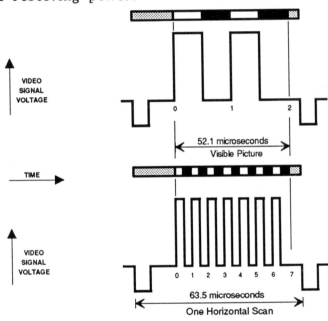

Figure 1.31
Video Signal Frequency Increases
when Number of
Black-to-White Transitions Increase

Vertical resolution is measured in a similar manner using horizontal converging lines on the resolution chart. Vertical resolution is inherently limited by the television scanning standards.

Notice that "lines of horizontal resolution" bears no direct correlation with the 525 horizontal "line" scan paths. On a 600-line resolution television monitor, there are 1200 brightness transitions visible on *each* of the 525 horizontal scan paths. As the horizontal scan paths are stacked one-on-top-of-the next, the white/black transitions *appear* to form vertical lines in the picture.

Another unit of measurement of picture resolution is the "pixel." The pixel represents the smallest possible discernible picture detail. Resolution of a system is usually expressed as the maximum number of discrete pixels that are generated or are visible in the horizontal and vertical axes of the display.

DYNAMIC RESPONSE

Dynamic response in pick-up devices is the ability to detect rapid changes in scene light intensity. Dynamic response problems frequently occur when a bright light source is replaced by a dark light source (like a light bulb that is turned from "on" to "off"). For many pick-up devices, this rapid light transition is difficult to accurately detect. Sometimes it takes several frames, or even several minutes of scanning to dissipate all the "burned-in" information in the target material. This produces a "ghost" or "lag" of the light scene in the dark picture for a while after the light scene has been removed. Moving targets in the frame may produce a smear of motion. Under severe burn-in conditions, the target material may be seared so that it never loses the bright details. In general, semiconductor-type pick-up tubes generally have much better lag characteristics than vidicons, particularly when "low-capacitive" target materials are used.

Another pick-up device problem becomes evident when there is a super bright object in an otherwise dark picture (such as sparkling highlights from jewelry or glasses under bright lights, etc.). The glint under some conditions may be too much for the pick-up device to handle. This "dynamic specular response" distortion is a particularly difficult problem that has been the subject of considerable research by pick-up device and camera manufacturers. Again, some types of devices have better characteristics than others. To optimize the response to these specular reflections, many camera manufacturers offer "automatic beam optimizer" (ABO) circuits and "diode gun" pick-up tubes that enhance response to specular reflections.

COLOR TELEVISION

The television system described so far can describe all the brightness details in a scene as a black-and-white picture, but no color. To understand color television, a review of color theory is in order.
Any color is made up of three component parts:

Luminance - overall (black-and-white) brightness

Hue - tint or "color"

Saturation - color intensity of the hue

The black-and-white television system describes only the luminance, forcing color television systems to describe hue and saturation by using special techniques.
Luminance, hue, and saturation of picture details can be approximated by unique combinations of primary color information. In television, the primary colors (corresponding to the subtractive color process) are:

RED

GREEN

BLUE

COLOR BARS

Figure 1.32 and 1.33 show how these primary colors can be combined to give the eight colors frequently seen in a color bar test pattern. (Split-field color bar "standards" are presented in Appendix E.) Notice that these eight colors are described by turning the primary color component parts either full "ON" or full "OFF" (like using a light switch). If other colors are needed, the intensity of the individual primary colors must be continuously variable from full "ON" to full "OFF" (like the effect of light dimmers on light output). In this way, nearly any color can be reproduced. Notice that Figures 1.32 and 1.33 also confirm two basic color theories; white is the equal presence of all colors, and black is the equal absence of all colors.

COLOR	RED	GREEN	BLUE
White	ON	ON	ON
Yellow	ON	ON	Off
Cyan	Off	ON	ON
Green	Off	ON	Off
Magenta	ON	Off	ON
Red	ON	Off	Off
Blue	Off	Off	ON
Black	Off	Off	Off

Figure 1.32
Color Bar Composition

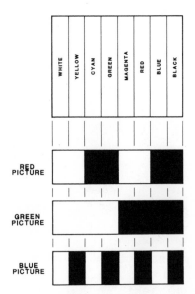

Figure 1.33
Red, Green, and Blue Pictures Generated by Color Bars

COLOR FOR REAL SCENES

This system of electronically generating colors works fine for color bars, but a single pick-up device detects the cumulative effect of *all* colors of light when focused on a real scene. Some method of enabling a pick-up device to detect the brightness of each primary color component of a scene must be provided.

It turns out that the brightness of an individual primary color component of a scene can be detected if an optical filter of the primary color is placed in the light path to the pick-up device as shown in Figure 1.34. The filter passes only the brightness information of the color of the filter and absorbs all the other colors. The action is similar to a brake light on an automobile, where white light is emitted from the light bulb and passes through the red filter to appear to shine red. The light energy of all the colors from the light bulb except red are absorbed by the filter and dissipated as heat.

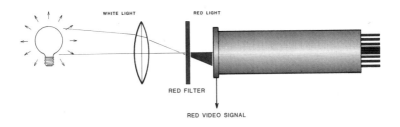

Figure 1.34
Red Filter Passes Only the Red Component of Light

Since brightness of the primary color can be detected with a filter and pick-up device, any color can be described by focusing the same scene through separate filters of the primary colors (red, green, and blue) on three separate pick-up devices. As shown in Figure 1.35, focusing the same image on all three pick-up devices may be accomplished with special "half-silvered" mirrors that allow only some hues of light to pass through each mirror while reflecting the remainder. In this way, one lens can be used to focus exactly the primary color components of the same scene on the target of each of the three pick-up devices.

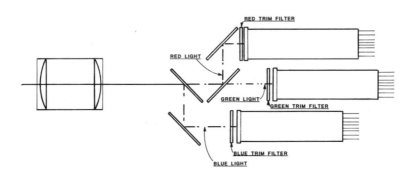

Figure 1.35
Typical Dichroic Mirror System

These "dichroic" ("split color") mirrors use special coatings to reflect specific primary colors and allow the remaining colors to pass. Even with dichroic mirrors providing color separation, the individual primary color paths usually have "trimming filters" to ensure accuracy of the primary color component detection.

To give an operating example, if a color camera is focused on a properly illuminated color "chip" chart, the video signal from each of the three pick-up devices is the same as the corresponding primary color video signals of the electronically generated color bar signal. For example, in magenta picture areas, the video signals from the red and blue pick-up devices are "high" and the video signal from the green pick-up device is "low;" in white areas the red, green, and blue video signals are "high;" etc.

As shown in Figure 1.36, primary color separation can also be accomplished by using a prism with dichroic surfaces. If a pick-up device is placed at each of the three exit angles that correspond to the three primary colors (red, green, and blue), they can detect the component colors of the light from the scene. The prism system *generally* allows more light to pass than the mirror system, making the camera more sensitive under given low light conditions. (This difference in light transmission is virtually negligible under normal operating conditions.) Usually, there are also fewer problems with chromatic aberrations and polarization effects with a prism system than through a mirror system.

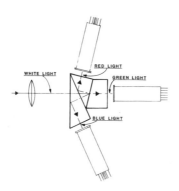

Figure 1.36
Typical Prism System

COLOR ENCODERS

Although the output of some systems are the component red, green, and blue video signals (R-G-B), in most instances they are combined (or "encoded") into a single video signal that is a composite of the primary color information. R-G-B goes into an "encoder" and the single encoded color signal comes out. The encoder may be a single integrated circuit in a camera or a separate piece of equipment with red, green, blue, and sync input connectors.

Countries in North America use color encoding standards established by the NTSC (National Television Systems Committee). Other countries often use different color encoding standards like PAL (Phase Alternation Line) or SECAM (Sequential with Memory). A television signal using NTSC encoding must undergo "transcoding" (conversion of the encoding process) for use in countries with other encoding or scanning standards.

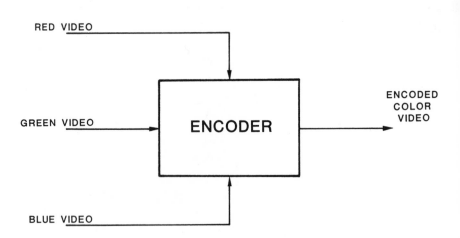

Figure 1.37
Encoder Combines Red, Green, and Blue
into a Single Signal

In the NTSC system, the luminance (black-and-white) signal is used as the electrical base on which the color signals are built. Color information in a scene is described by varying the timing and electrical power of a special "subcarrier" signal. As shown in Figure 1.38, when this subcarrier signal is added to the luminance signal, any color can be completely described.

Early color television cameras used four pick-up tubes - one for each of the three primary colors and one without any optical filtering for the luminance information. After years of refinement, it was discovered that acceptable results could be obtained from three tubes by designing the encoder to artificially generate a close approximation of the true

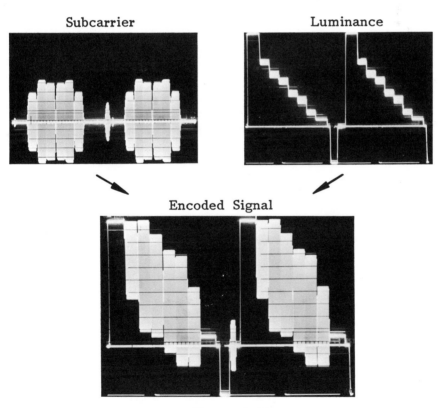

Subcarrier   Luminance

Encoded Signal

Figure 1.38
Adding Subcarrier to Luminance

luminance signal by combining the separate red, green, and blue video signals in the following way:

30% of the Red Video Level

PLUS

Luminance Level =   59% of Green Video Level

PLUS

11% of Blue Video Level

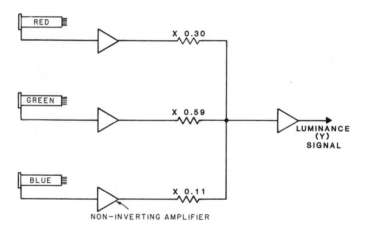

Figure 1.39
Luminance Signal Development

Once the luminance signal is generated, the saturation and hue information need to be added. To generate the subcarrier signal that accurately describes the saturation and hue of picture detail, two additional combinations of the red, green, and blue video signals are derived.

One of these signals, the "I" modulating signal (I = "in phase"), consists of the following video signal components:

60% of Red Video Level

MINUS

I Modulating Level =   28% of Green Video Level

MINUS

32% of Blue Video Level

(Signals are subtracted by turning the subtracted signal upside down, or "inverting," where white becomes black and black becomes white before being combined with the other signals.)

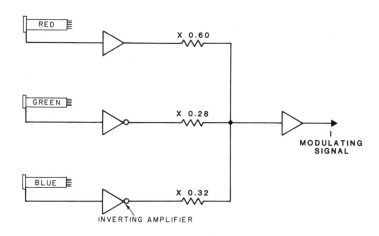

Figure 1.40
I Signal Development

The second of these signals, the "Q" modulating signal (Q = "quadrature phase"), consists of the following video signal components:

21% of Red Video Level

MINUS

Q Modulating Level = 52% of Green Video Level

PLUS

31% of Blue Video Level

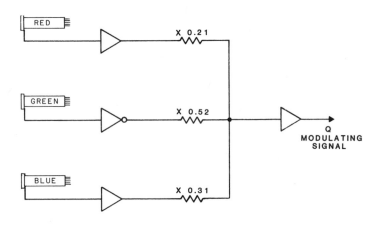

Figure 1.41
Q Signal Development

The I and Q modulating signals are then sent to the I and Q modulators to reduce the visible effects of distortions in the signal. As shown in Figure 1.42, these modulators vary the amplitude (peak-to-peak voltage) of the color subcarrier in response to the voltage variations of the modulating signal. The output of the I modulator is a signal with the color subcarrier frequency (3,579,545 Hz) that has amplitude variations corresponding to the amplitude variations of the I modulating signal; likewise, the output of the Q modulator is a signal with the color subcarrier frequency that has amplitude variations corresponding to the amplitude variations of the Q modulating signal. Other than the amplitude variations caused by the different modulating signals, there is one other difference between the two signals - the timing of the peaks and valleys of the subcarrier signal base.

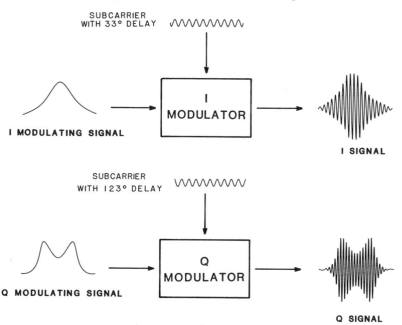

Figure 1.42
I and Q Modulators

If two subcarrier signals **both with 3,579,545 cycles-per-second** are examined, their peaks and valleys *may* occur at exactly the same time. When this happens, as shown between signals A and B in Figure 1.43, there is no timing difference and the signals are said to be "in phase." But if one signal is delayed and the other signal is left alone, the peaks and valleys do not occur at the same time, changing the phase relationship between the two signals. *The frequency of both signals has not changed from 3,579,545 cycles-per-second, just the relative timing between the two signals.* (Actually there is a small instantaneous frequency change whenever a phase change occurs, but this frequency change is transient and small.)

51

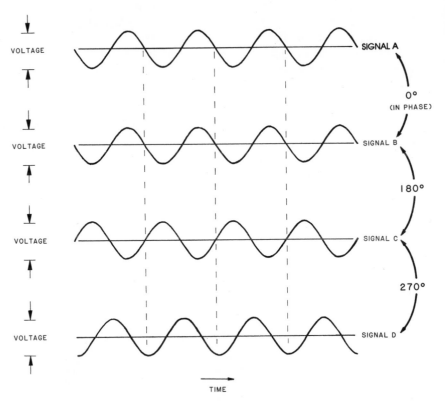

Figure 1.43
Phase Relationships of Signals

Phase relationships are expressed in "degrees" and there are 360° in each cycle of signal. (This is explored in detail when vectorscopes are discussed.)

As shown in Figure 1.44, the subcarrier signal that is input to the encoder (through a Subcarrier Input connector on the back of a camera control unit or a stand-alone encoder) is delayed by 33° before being fed to the I modulator. Thus the output of the I modulator is a signal of subcarrier frequency, 33° out of phase relative to the input subcarrier, and has amplitude variations that correspond to the variations in the I modulating signal. Before going on to the Q modulator, the subcarrier is then delayed an *additional* 90°. Thus the output of the Q modulator is a signal of subcarrier frequency, 33°+90°=123° out of phase relative to the input subcarrier, and has amplitude variations that correspond to the variations in the Q modulating signal.

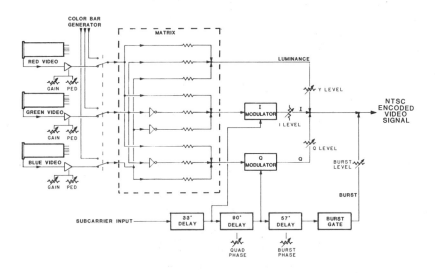

Figure 1.44
NTSC Encoding Block Diagram

Because of the 90° phase difference between the I and Q signals, each unique *combination* of I and Q signals can describe the amplitude and phase of a signal that can be "decoded" to display a particular hue and saturation. (More detail is be provided in later discussions about color signal analysis.)

Examination of a real encoded signal shows that the subcarrier phase, the subcarrier amplitude, and the luminance amplitude are constantly changing as each detail in a focused scene is detected and encoded. In the midst of all this radically changing electrical information, a constant phase reference must be provided to allow accuracy in the signal decoding processes. To provide this reference, the scene information from each horizontal scan is preceded by a "burst" of several cycles of subcarrier with a constant phase. (Detailed use of the burst is explored in later color signal analysis discussions.)

The luminance components provide an electrical base to which the color components are added. In this way, compatibility is maintained between black-and-white systems (which use only the luminance after filtering out all the subcarrier frequency signals) and color systems (which use all the components).

53

```
┌─────────────────────────────────┐
│          LUMINANCE              │
│                                 │
│            plus                 │
│                                 │
│          I SIGNAL               │
│      (modulated subcarrier)     │
│                                 │
│            plus                 │
│                                 │
│          Q SIGNAL               │
│      (modulated subcarrier)     │
│                                 │
│            plus                 │
│                                 │
│        BURST SIGNAL             │
│      (gated subcarrier)         │
└─────────────────────────────────┘
```

Figure 1.45
NTSC Signal Composition

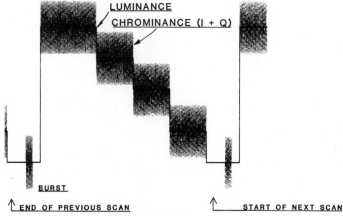

LUMINANCE
CHROMINANCE (I + Q)

BURST
↑ END OF PREVIOUS SCAN   ↑   START OF NEXT SCAN

Figure 1.46
Video and Blanking (VB) Voltage Graph
of an NTSC Encoded Signal

Figure 1.44 is a block diagram of a typical NTSC encoder and can be examined to give an overview of how the encoded color signal is generated. The drawing also details some of the adjustments that can be anticipated in most encoder circuits. Proper adjustments of an encoder require an oscilloscope (or waveform monitor) and a vectorscope. Procedures for encoder circuit adjustment of a particular piece of equipment may be found in the appropriate service manual.

In the NTSC system, the resultant hue and saturation of a color are described as the result of combining the I signal with the Q signal (in a way to be described in the color signal analysis section). As shown in Figure 1.47, picture hue is described by comparing the phase of the resulting subcarrier with the burst. As shown in Figure 1.48, picture saturation is described by the amplitude of the resulting subcarrier.

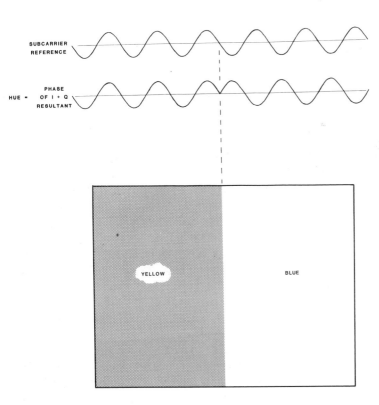

Figure 1.47
Chrominance Phase Determines Picture Hue

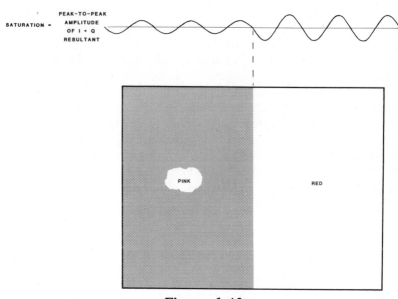

Figure 1.48
Chrominance Amplitude Determines Color Saturation

Some "NTSC-type" encoding schemes do not use the common 33° phase shift between the subcarrier provided the modulators and the system subcarrier. The "R-Y Modulator" is provided subcarrier at 180° phase and the "B-Y Modulator" is provided subcarrier at 90° phase relative to the burst and system subcarrier. The "Red Minus Luminance" (R-Y) and "Blue Minus Luminance" (B-Y) modulating signals are easier to generate than the I and Q matrixing schemes, leaving green to be derived algebraically. "Component" video tape systems use the simplified matrixed red, green, and blue signals combined as R-Y and B-Y.

The chrominance and the luminance make up the analog component parts of a color television signal. By keeping these component parts separated, the interaction between chrominance and luminance that produces picture distortions in an NTSC encoded signal is minimized. As explored later, keeping the chrominance and luminance components separated can dramatically improve picture quality.

ONE-DEVICE COLOR CAMERAS

The three pick-up device camera is straightforward in the theory that enables it to separate scene colors into their primary color component parts. But techniques have been developed to enable a single pick-up device to detect scene colors.

The scanning process happens so fast that acceptable color pictures can be obtained if a single video signal is "time shared" to approximate the three primary color video signals. Many variations on the one-device technique are used, but in general the faceplate of the single pick-up device is patterned with combinations of optical filters of primary or complementary colors. Figure 1.49 shows a pick-up tube from a one-tube color camera. The filters are so small that they cannot be seen in this photograph. As the scan goes from left-to-right and top-to-bottom across this tube's target, the tube momentarily reads filtered brightness information corresponding to each filter color. The metallic areas in the target area of the photograph are indexing electrodes to positively identify which filter is being used at any given time.

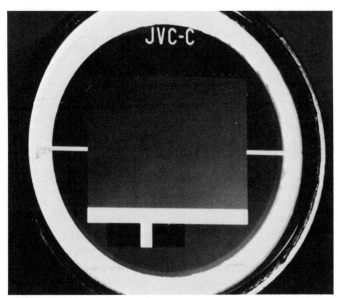

Figure 1.49
Faceplate of a One-Tube Color Camera's Pick-up Tube

As shown in Figure 1.50, the signals from each of the filtered areas is temporarily memorized so that brightness information of an individual color is stored and used during the readings of the other colors. A fast "electronic switch" determines the time during which the single video signal is interpreted as representing a particular color. Because of the temporary storage, the video signals appear to be continuously output until the memory can be refreshed with information from the next point on the target where light passing through the same filter color is detected. These memorized and refreshed signals are then fed into the encoder for processing.

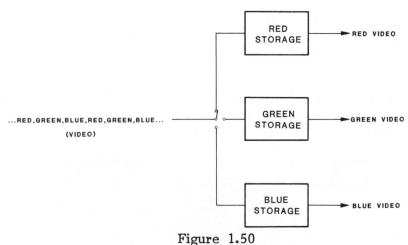

Figure 1.50
Video Switching and Storage
in a Typical One-Tube Color Camera

One-device cameras which use primary color filter separation use essentially the same type of encoder as a three-device camera. If the individual color filters use a combination of primary colors, the encoder and electronic switch must be redesigned to recognize these combinations for use in developing the I and Q or the R-Y and B-Y subcarrier modulating signals.

The video signals of the component colors from one-device camera circuits appear to be continuous, thanks to the temporary storage. This storage holds the color information until it can be updated with brightness information from the next point on the target where light is passing through the same filter color. The component color signals are then fed into an encoder for processing.

As might be expected, the one-device color television camera offers significant economic advantages over the three-device camera. But there is usually a significant decrease in performance. Basically, the one-device camera provides one-third the color resolving capability of a three-device camera, making sharp color transitions difficult to detect. If one were to have a scene that has a sharp vertical black-to-white transition as shown in Figure 1.51, it may be quite a while before the transition is detected in all three color signals. Using a device that has vertical primary color striped filters as our

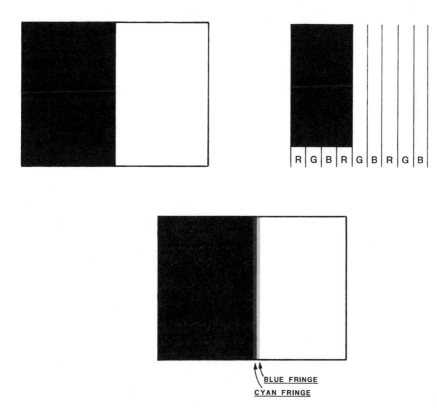

Figure 1.51
Color Fringing in a Typical One-Tube Color Camera

example, assume the black-to-white transition happens to fall between a red and an adjacent green filter stripe. The red filter sees black, the adjacent green filter sees white, and the following blue filter sees white. It is not until the following red filter is scanned that the red video channel perceives white like the other two video channels. This delay would result in a small green + blue = cyan fringe on the right side (latest scanned side) of the black-to-white transition. Once the red video channel finally gets the information that a white picture is now present, the red + green + blue = white picture will be produced. Manufacturers have developed various techniques of using super-thin filter stripes, diagonal filter stripes and filters of combinations of primary colors to reduce the size of the areas of the picture where this color fringing effect is objectionable.

PICK-UP TUBE OPERATING CONTROLS

There are four basic electronic operating controls common to all pick-up tube cameras. Adjustment of TARGET, ELECTRICAL FOCUS, BEAM and GAIN controls determine the camera's picture quality by changing the electric fields within the pick-up tube and through its associated circuitry.

The operation of the pick-up tube controls can best be understood if electron reaction to outside influences is first examined. From previous discussions about electron beam deflection for scanning the target, there are three basic points to remember about electron behavior:

1) Electrons are *negatively-charged* particles.

2) An electron tries to get farther *away from another negative charge.* (Because "like" charges repel.)

3) An electron tries to get closer *to a positive charge.* (Because "opposite" charges attract.)

TARGET CONTROL

To coerce the electrons to travel from the electron generating "cathode" (at the back of the pick-up tube) to the target (at the front of the tube), they must "see" a positive charge in the direction of the target.

The TARGET control shown in Figure 1.52 determines the amount of positive charge applied to the target to optimize operation between two extremes:

If the TARGET control is adjusted so that the charge on the target material is *too positive*, the electrons are attracted too strongly, hitting the target material so hard that they bounce off and land on adjacent target areas, reducing picture detail. Since electrons are particles resembling small sand grains, they may eventually erode the target material until the tube is unusable. Pick-up tubes are expensive, so the slowest erosion of target material is desirable.

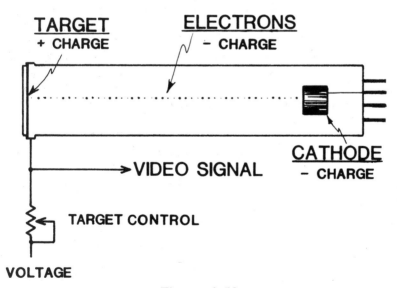

Figure 1.52
TARGET Controls the Positive Charge on the Target

If the target is *not positive enough*, it can't attract enough electrons to accurately describe all the picture details. If too few electrons are attracted to flow through the target, even the smallest target electrical resistance may be too high to allow enough electrons through to describe any of the picture details. Even if some of the picture details are present, too low a charge on the target can allow the picture to lose detail in the white areas or "lag" with a trailing smear of any moving picture details. Taken to extreme, the charge on the target may be made so negative that the electrical current reverses to invert the video signal and create a negative picture.

The TARGET control should be adjusted to the lowest possible setting where all the white picture details are visible.

With semiconductor-type pick-up tubes (like the Plumbicon®, Saticon®, etc.), the TARGET control has a less drastic effect on the finished picture because it only changes the operating point, or "bias," of the target material between the extremes of complete cut-off of signal flow and the destruction of the target material. With any pick-up tube, the TARGET control should be adjusted while referencing to the camera's *service* manual and pick-up tube specifications. Figure 1.24 revealed a target (signal electrode) voltage of 45 volts relative to the cathode voltage. Figure 1.53 shows some typical target voltage ranges.

TARGET VOLTAGE (relative to cathode)	
VIDICON	10 - 60 Volts
PLUMBICON®	40 - 45 Volts
SATICON®	45 - 50 Volts

CHECK THE TUBE SPECIFICATION SHEET!
Figure 1.53
Typical Target Voltages

ELECTRICAL FOCUS CONTROL

The ELECTRICAL FOCUS control determines the *cross-sectional size* of the electron beam when it reaches the target. The smaller the size of the beam, the more picture detail can be resolved. To maximize picture resolution, the electron beam is shot down the long axis of a negatively-charged cylinder within the pick-up tube (and frequently through the electric field created by an electromagnet encircling the tube). The effect of these combined electric fields forces the stream of negatively-charged electrons into the smallest possible cross-sectional area. (*ELECTRICAL FOCUS adjustments for the pick-up tube should not be confused with any of the optical focus adjustments associated with a lens - both the light and the electron beam must be focused for the sharpest possible picture.*)

If the cylinder of electric charge is *not negative enough*, the electrons are not compressed into the smallest possible size. The cross-sectional area covered by the stream of electrons may be so large that it covers adjacent details in the image focused, on the target, reducing picture sharpness and detail resolution.

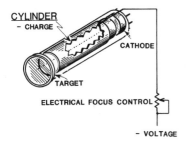 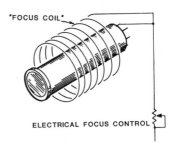

Figure 1.54
Electrical FOCUS Control Adjusts Electric Field
Created by the Focus Cylinder or Coil

If the cylinder of electric charge is *too negative*, any slight deformation in the cylinder or the spin of the individual electrons forces some of the electrons off-axis to enlarge the size of the beam, reducing picture sharpness.

A rotation of the image is often observed when adjusting electrical focus. This rotation is a result of the interaction of the electric fields being generated by the focus coil (or tube element) and the tube's deflection coils (or plates). On many cameras, this rotation is adjustable to be in the center of the picture to assure balanced electrical focusing capability across the entire scanned area.

Most tube-type cameras are equipped with "dynamic focus" circuits to change the strength of the electric field generated by the focus coil (or tube element) at a rate determined by the scanning rate. Dynamic focus circuits allow generation of pictures with optimized sharpness near the edge of the scanned area. The dynamic focus circuits force the electron beam to contact the target material at a 90° angle so that the effective shape of the beam is circular. Without dynamic focus circuits, the electron beam strikes the target material at an angle and create an elliptical cross-sectional shape (in the same way that a circular spotlight beam becomes elliptical when it hits the stage floor at an angle). The enlarged elliptical shape of electron beam reduces picture resolution.

To aid in the adjustment of dynamic focus, many cameras offer a "focus wobble" mode to rapidly vary the focus. Proper adjustment of focus is achieved when the picture wobbles or rotates about the exact center of the scanned frame.

ELECTRICAL FOCUS should be adjusted for the best picture detail and sharpness.

BEAM CONTROL

The BEAM control limits the *number of electrons* that are available in the pick-up tube to form the electron beam. The cathode generates many, many more electrons than are needed. To reduce and control the number of electrons, a negatively-charged wire screen is placed in the path of the electrons on their way to the target. Some electrons get through the screen, but most are repelled back.

If the charge on the wire screen is *too negative*, too few electrons can get through to form an electron beam of sufficient strength. If too few electrons pass through the screen, there is insufficient electrical flow to describe all of the picture details (starting with loss of detail in the white areas). The TARGET and BEAM controls interact to determine the optimum operating point for a given pick-up tube.

If the charge on the wire screen is *not negative enough*, too many electrons are allowed to pass through to be squeezed into a beam of the smallest cross-sectional area, reducing picture sharpness.

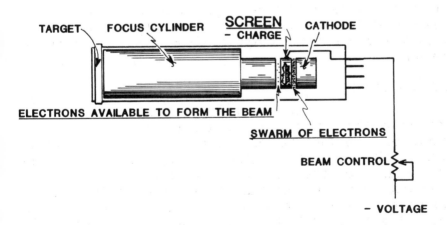

Figure 1.55
BEAM Controls the Negative Charge
on a Screen in the Pick-up Tube

The BEAM control should be adjusted to the minimum setting where the white picture details are visible.

Some cameras have an automatic control of the beam current to allow instantaneous changes that are based on picture content. This "automatic beam control" or "beam optimizer" capability allows what is essentially a greater contrast range. Super bright details in a bright picture can be more accurately described, reducing the effects of "flare" and "comet-tailing."

With the BEAM, TARGET, and ELECTRICAL FOCUS controls properly set, the pick-up tube converts a visual scene into a video signal. Each pick-up tube has slightly different requirements for proper operation, so a three-tube camera has three BEAM controls (one for the red tube, one for the green tube, and one for the blue tube), three TARGET controls, and three ELECTRICAL FOCUS controls. On many camera models, manufacturers do not recommend frequent user adjustment of the BEAM, TARGET, or ELECTRICAL FOCUS controls (as indicated by controls that require a screwdriver for adjustment).

In cameras equipped to use semiconductor-type pick-up tubes (like the Plumbicon®, and Saticon®), a small lamp or photodiode is designed to shine evenly across the target material of each tube. This "bias light" minimizes lag and smearing of moving images in dimly lit scenes by altering the response of the target material. In color cameras, bias light adjustments are provided for each tube. Adjustment of bias light intensity changes the color balance of a multiple-tube camera.

GAIN CONTROL

The video signal generated by the pick-up tube is so weak that it must be amplified (made more powerful) to be usable. The amount of strength that the signal gains as it goes through the camera is determined by the GAIN and BOOST controls. On most cameras there are several GAIN controls inside the camera case with a switch on the outside for the operator to BOOST the gain "-4," "0," "+6," "+9," "+12," "+18," etc. The numbers refer to the number of deciBels (dB) that the signal is changed in strength. More about dBs in a later discussion.

66

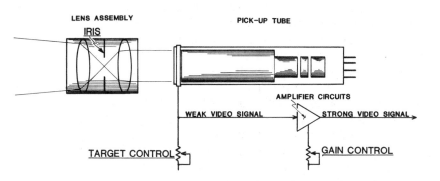

Figure 1.56
GAIN Controls the Amount of Power
Gained by the Video Signal

The amount of light passing through the lens, the TARGET control, and the GAIN control interact to determine the contrast of a picture. Proper adjustment of the camera requires a specified amount of light *reaching the pick-up tube*, the TARGET control properly adjusted; the BOOST control set to "0;" and the various GAIN controls adjusted for a camera output of 0.714 Volts (peak-to-peak) of video signal (picture only). Intermediate GAIN controls are usually provided for each channel (R-G-B) in a three-tube camera as well as a master GAIN control for adjustment of the peak-to-peak voltage of the output color signal.

The camera's BOOST control allows the operator to compensate for low light. In the "0 dB" position, the camera is using just enough gain of the video signal voltage to give 0.714 volts output under the specified lighting level and the specified ratio of desired signal voltage to the undesired random noise signal voltage. Should the camera need to be operated under lower scene illumination levels, additional gain can be switched in by switching the BOOST control to a positive number.

As seen in Figure 1.57, when the BOOST control is switched to something greater than 0 dB, the signal-to-noise ratio is decreased. The +6 dB BOOST setting doubles the desired signal voltage and doubles the undesired noise voltage *plus* adds the noise created in the additional amplifier circuitry. The added amplifier noise reduces the signal-to-noise ratio to increase the amount of random noise "snow" present in the picture.

Some cameras also have a negative BOOST switch position to improve the signal-to-noise ratio under very bright scene lighting intensities.

The BOOST control can be used by the operator to creative advantage. For example, if a large depth of field is needed to keep two separated objects in focus, the iris opening may be reduced by one f-stop by increasing the gain. Basically, the amount of light falling on the scene can be doubled (to keep the same amount of light reaching the pick-up tube) or the BOOST can be switched for 6 dB additional gain. The depth of field is increased at the expense of signal-to-noise ratio.

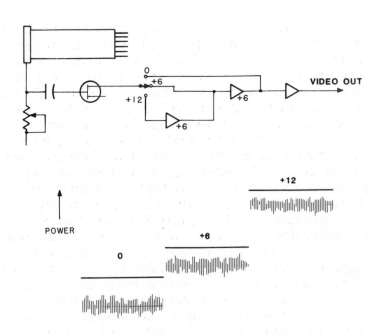

Figure 1.57
Increasing Gain Decreases Signal-to-Noise Ratio

COLOR TELEVISION CAMERA OPERATING CONTROLS

REGISTRATION CONTROLS

With more than one tube in a color television camera, the focused image on the target of each pick-up tube must be scanned in *exactly* the same way as the remaining tubes. The easiest way to visualize this requirement is to imagine a grid of horizontal and vertical white lines on a black background. As shown in Figure 1.58, if the three pick-up tubes do not scan the grid in exactly the same way, two or more separate grids (one from the red, one from green, and one blue) may be seen in the finished picture. To eliminate this problem, the output picture of the three tubes must be "registered" to appear superimposed atop of each other.

Mechanical adjustments are provided for a technician to adjust coarse registration of the tubes in a multiple-tube color television camera. For "fine tuning" registration, the operator has electrical controls for adjustment of the deflection of the electron beam (and the relative location of the scan on the target material) in each pick-up tube.

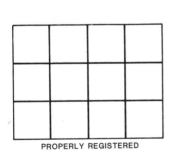

PROPERLY REGISTERED

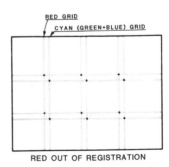

RED OUT OF REGISTRATION

Figure 1.58
Registration

69

Most cameras use the green tube as the reference for optical coincidence of the other tube(s) so there are a minimum of four registration controls available to the operator on a three-tube camera (the green tube has the same four control functions, but they are usually limited to technician use):

RED Vertical Centering

RED Horizontal Centering

BLUE Vertical Centering

BLUE Horizontal Centering

A test chart with a black grid on a white background similar to that shown in Figure 1.59 is normally used for easy adjustment of the registration controls. As shown in Figure 1.60, the green video signal is inverted to form a white grid on a black background. When added to either the red or the blue video signal and registration is perfectly adjusted, a solid gray picture results with no grid readily apparent. (The white details of the negative green and the black details of the red or blue "null" to form gray details. The black details of the negative green and the white details of the red or blue combine to form details with the same gray brightness.) To see the finest possible details in the picture while adjusting registration, a high resolution black-and-white picture monitor is normally used.

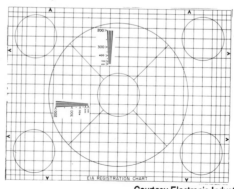

Courtesy Electronic Industries Association

Figure 1.59
Registration Chart

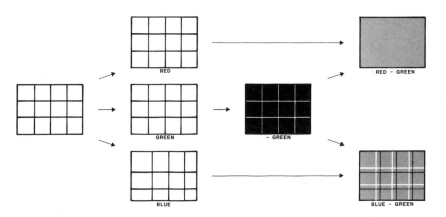

Figure 1.60
Registration Procedures
(Blue Out of Registration)

Besides the four operator controls, there are usually several controls available inside the case for the technician to adjust the scanning of each tube for a minimum of registration error. As shown in Figure 1.61, each of these controls can correct for error or can introduce error. When these controls are improperly adjusted, the center of the picture can be in correct registration (by the operator using the centering controls), but the edges and corners may still be out of registration. Automatic centering or shift circuits adjust only the vertical and horizontal centering for overlay at picture center.

It's easy to visualize the effect on the picture from an individual tube when the registration controls are varied if the voltage ramps applied to the deflection coils (or plates) are examined. The centering controls adjust the voltages of the starting and stopping points of the voltage ramps equally. Looking at Figure 1.62, controls labeled AMPLITUDE, SIZE, WIDTH, and HEIGHT adjust the voltage difference between the peak-to-peak voltage of the ramps. LINEARITY controls adjust the voltage of either the start or stop points relative to the mid-point of the ramp. LINEARITY and SIZE controls usually interact.

71

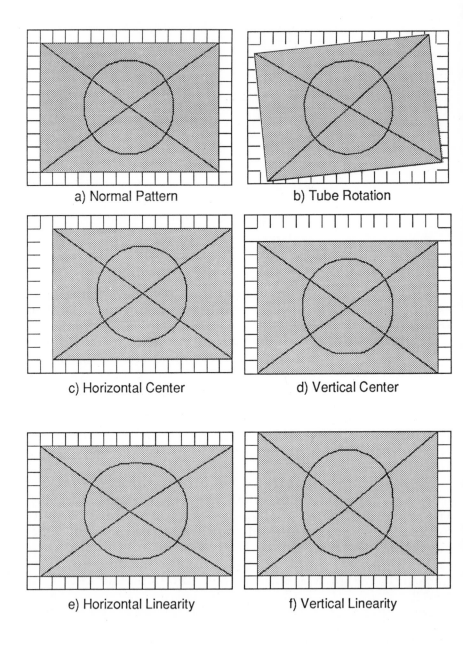

a) Normal Pattern b) Tube Rotation

c) Horizontal Center d) Vertical Center

e) Horizontal Linearity f) Vertical Linearity

Figure 1.61
Effect of Registration Controls on the Picture

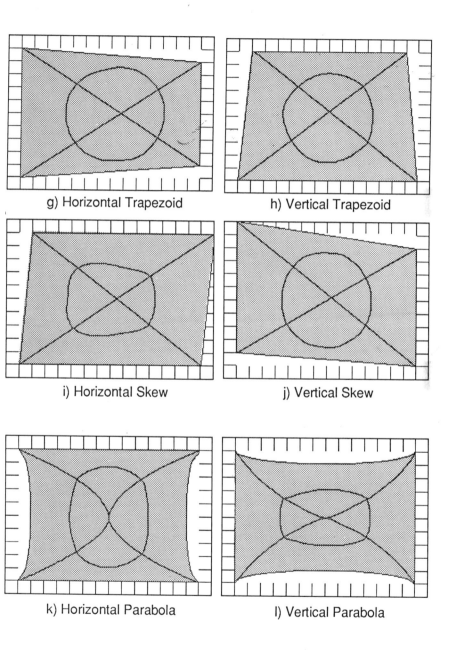

g) Horizontal Trapezoid

h) Vertical Trapezoid

i) Horizontal Skew

j) Vertical Skew

k) Horizontal Parabola

l) Vertical Parabola

Figure 1.61 *(cont.)*
Effect of Registration Controls on the Picture

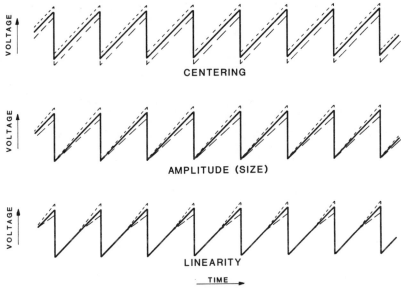

Figure 1.62
Deflection Voltage Changes
From Registration Adjustments

Multiple-tube camera registration should be checked before each use and frequently rechecked during production. Modern camera designs that incorporate circuits to reduce drift in the pick-up tube deflection circuitry have reduced the frequency of adjustment under normal operating conditions, but any drastic change in the operating temperature or any mechanical or electrical shock to a camera should be followed with a check of the registration status.

COLOR BALANCE

Another control that is common to all color television cameras adjusts the camera to detect the proper colors with variations in the light illuminating the scene. Every light source has a different "color temperature." Incandescent bulbs produce a yellowish light, fluorescent bulbs frequently produce a greenish light, and sunlight varies according to the time of day, the season of the year, and the weather. A camera must alter its response to various colors to compensate for differences between colorimetric values sensed by the camera and the human eye.

The color temperature of a light source is expressed in Kelvin and abbreviated as K. (Many sources still cite the old unit of "degrees Kelvin," or "°K.") This is the temperature to which a perfect black body in a perfect vacuum must be raised to generate a spectrum of light closest to the light source color (like raising the temperature of a metal through a red glow to "white hot"). Note that this is a theoretical definition, for neither a perfect black body nor a perfect vacuum really exists.

Most television cameras are optimized for operation with quartz iodine lighting of 3200K and use optical filters to coarsely adjust other color temperatures within the electronic compensating limits of color balance controls in the camera. Common filters and the intended light sources are:

3000K to 3200K : Incandescent Light

4500K to 4800K : Cloudy or Rainy Daylight

5900K to 6500K : Bright Daylight

Fluorescent lighting produces a limited spectrum of colors that cannot be accurately described in terms of Kelvin. It's difficult (and sometimes impossible) to compensate for accurate color rendition of a scene illuminated by normal fluorescent lamps. Under such conditions, the area normally illuminated by fluorescent sources is usually bathed with acceptable light to minimize the greenish color normally produced by fluorescent sources.

The human brain has an automatic color balance adjustment that operates so fast that we don't notice the subtle changes of color. But the television camera must be told by the operator exactly which of the picture areas should be used as a reference to adjust for these variances.

White is composed of all the colors in equal amounts, so it is the best reference to use for adjustment of camera response to various color temperature light sources. WHITE BALANCE (or COLOR BALANCE) controls adjust the electrical response of the television camera so that white is interpreted and displayed as white to the eye, regardless of any color contamination that is present in the incident light source.

To see how this is done, assume 0.714 Volts in *each* primary color video channel indicates white, remembering that white contains equal amounts of each primary color. Then, using the formulas from the encoder discussions:

$$Y = (.30 \text{ X Red}) + (.59 \text{ X Green}) + (.11 \text{ X Blue})$$
$$= (.30 \text{ X } .714) + (.59 \text{ X } .714) + (.11 \text{ X } .714)$$
$$= 0.714 \text{ Volts}$$

$$I = (.60 \text{ X Red}) - (.28 \text{ X Green}) - (.32 \text{ X Blue})$$
$$= (.60 \text{ X } .714) - (.28 \text{ X } .714) - (.32 \text{ X } .714)$$
$$= 0 \text{ Volts}$$

$$Q = (.21 \text{ X Red}) - (.52 \text{ X Green}) + (.31 \text{ X Blue})$$
$$= (.21 \text{ X } .714) - (.52 \text{ X } .714) + (.31 \text{ X } .714)$$
$$= 0 \text{ Volts}$$

A pure white is assured when the gain of each of the video channels is adjusted (using WHITE BALANCE) so that the encoder produces no color subcarrier to be added to the luminance in the white picture areas. Figure 1.63 shows that the signal may be graphically examined using a waveform monitor for presence of color-contaminating subcarrier in white picture areas. (A detailed discussion about waveform monitors is presented later.)

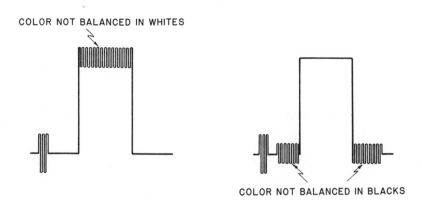

COLOR NOT BALANCED IN WHITES

COLOR NOT BALANCED IN BLACKS

Figure 1.63
White Balance and Black Balance
Adjusts for Minimum Subcarrier

Many cameras now feature automatic adjustment of white balance to adjust the gain in the red, green, and blue channels for elimination of the color subcarrier from the white picture areas designated by the camera operator. The operator usually focuses what he wants to designate as white in a particular area of the scanned picture, then momentarily presses a button to activate the camera's automatic white balance procedure.

Some cameras also have operator controls to adjust the purity of the "black" base on which the pure "white" picture can be built. In cameras with manual controls available to the operator, the BLACK BALANCE (or red, green, and blue PEDESTAL) control(s) should be adjusted to give a properly balanced base on which the pure "white" picture can be generated. As shown in Figure 1.63, when color balance is properly adjusted, the encoder generates no subcarrier on the luminance base when viewing black, white, or gray picture details.

IMAGE ENHANCEMENT

Many cameras have circuits called image enhancement, aperture correction, contour, or detail. These circuits tend to sharpen picture detail by artificially sharpening the brightness transitions in the picture. The block diagram of one circuit design for sharpening picture transitions is shown in Figure 1.64. This circuit compares and averages picture information on horizontal scans adjacent to the current horizontal scan by delaying the previous scan by 2H (two times the time spent in a horizontal scan), delaying the current scan by 1H, and introducing no delay in the future scan. In this way, the past, current, and future scans arrive at a single point and a single time to allow instantaneous averaging.

Like the delay of horizontal scan information, to enhance the sharpness of the picture on the vertical axis, horizontal axis sharpness is treated in 100 nanosecond (0.0000001 second) increments. This time correlates to the nominal size of the smallest visible horizontal picture element produced by this particular camera.

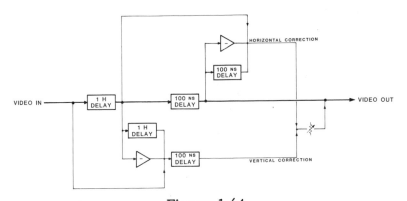

Figure 1.64
Typical Image Enhancement Circuit Block Diagram

A control is available to determine the amount of correction that is added to the original signal. An understanding of proper adjustment of the DETAIL control is apparent if the generated signals are examined. As shown in Figure 1.65, the signal comes from a pick-up tube with the sharp transitions in the real scene rounded off because of distortions created by the finite "aperture" size. Without correction, the picture tends to be fuzzy and indistinct. This is caused by the inability of the entire cross-section of the electron beam to instantaneously sense a sharp picture transition.

Where the video signal is too low before the sharp transition, the aperture correction signal adds voltage. As can be seen before the transition, at the mid-point, and after the transition; where the video signal needs no correction, no error correction voltage is created. At the end of the transition, a negative correction voltage is generated.

If the correction voltage is of the proper amplitude, the correction signal added to the aperture distorted video signal should approximate the original scene. If too much correction is added, white and black fringes of the transition is evident in the picture. If too little correction is added, the picture is indistinct.

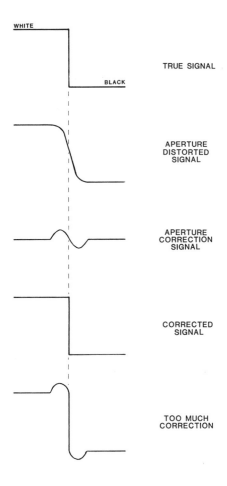

Figure 1.65
Detail Correction Signals

Detail circuits may also enhance any noise in the the picture to reduce the apparent signal-to-noise ratio. To reduce this signal degradation, many cameras switch the detail circuits off when using gain BOOST. Other cameras use a dynamic detail adjustment circuit that adjusts the amount of correction depending on scene brilliance or when noisy picture conditions exits.

In a multiple-tube color camera, DETAIL should be turned off when registering the camera. This eliminates any confusion between erroneous detail corrections signal detail and registration error.

COLOR CAMERA BLOCK DIAGRAM

For reference, the block diagram for a typical three-tube color television camera is shown in Figure 1.66.

CAMERA CONTROL UNITS

Some camera configurations allow a separated camera "head" to be remotely controlled from a distant "camera control unit" or CCU. A CCU controls critical technical operation of the camera, allowing operations where one person concentrates only on camera aiming for scene framing and another person continually monitors and adjusts electronic camera controls for the best possible picture.

CCUs provide remote control of registration, manual or automatic iris, master pedestal, white balance, black balance, and other controls that may be used in the course of production. Remote Control Units (RCU) or Operation Panels (OP) may be used to provide remote control of a smaller number of adjustments. As shown in Figure 1.67, interconnecting the CCU and the camera head is usually a heavy multi-conductor cable that conveys video, sync, power, intercom, tally, audio, and remote control signals.

Other cameras may use a special "triaxial" cable to interconnect a CCU "base station" with the camera head to convey the required control, production, and program signals. Triax cable is usually lighter and allows operation over greater distances than conventional camera cable. On the negative side, triaxial systems usually cost more than conventional systems.

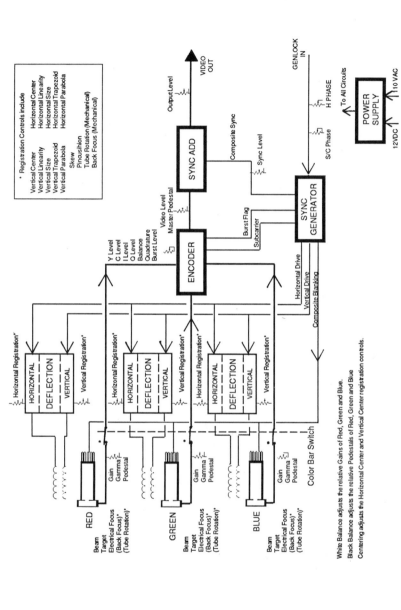

Figure 1.66
Typical Color Television Camera Block Diagram

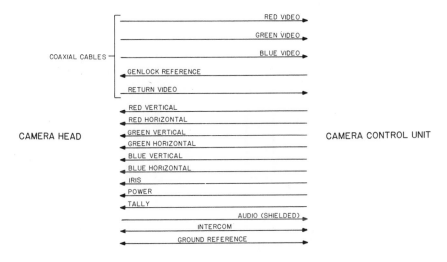

COAXIAL CABLES

RED VIDEO
GREEN VIDEO
BLUE VIDEO
GENLOCK REFERENCE
RETURN VIDEO

CAMERA HEAD

RED VERTICAL
RED HORIZONTAL
GREEN VERTICAL
GREEN HORIZONTAL
BLUE VERTICAL
BLUE HORIZONTAL
IRIS
POWER
TALLY
AUDIO (SHIELDED)
INTERCOM
GROUND REFERENCE

CAMERA CONTROL UNIT

Figure 1.67
Typical Signals Between Camera Head and CCU

Figure 1.68
Typical Camera Using Triaxial Cable
Between Camera Head and CCU

COLOR VIDEO MONITORS

To convert the video signal back into visual information, a video monitor must put the picture back together in exactly the same pattern from which the scene was detected by a camera.

The predominant part of the video monitor is the picture tube. In this glass envelope, shown in Figure 1.69, a beam of electrons is shot from a cathode (at the rear of the tube) to hit a coating of special material at the front (viewing end) of the tube. When the electron beam strikes this coating of "phosphors" at the front of the tube, they glow through the glass tube in proportion to the intensity of the beam. By placing a deflection yoke similar to that shown in Figure 1.69 with electromagnets at the top, bottom, left, and right of the tube and applying varying voltages to deflect the electron beam, scanning is accomplished to create a "raster" of energized phosphors for video signal display.

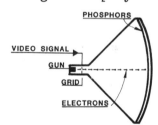

a) Side View

b) Rear View

Figure 1.69
Typical Picture Tube

The intensity of the electron beam is controlled by shooting the beam through a screen that has a negative charge (in exactly the same fashion as the grid in the pick-up tubes). The amount of charge on the screen is varied by the video signal, varying the intensity of the electron beam striking the phosphors in proportion to the original scene. A relatively high voltage in the video signal corresponds to white picture detail and commands the monitor to generate an intense electron beam that strikes the phosphors and makes them glow brightly. Conversely, when the picture detail in the video signal is dark, the monitor is commanded to generate a weaker electron beam that strikes the phosphors and makes them glow dimly (if at all). Phosphors for black-and-white video monitors are designed to glow white when energized and remain black when not energized.

When color is desired, the encoded color signal input to the monitor is first separated into its luminance and chrominance components. (If the input to the monitor is separated red, green, and blue video signals, this step is omitted.) Since the 3.58 MHz color information is buried within the luminance information (which frequently has components up to 4.2 MHz), the filter may be designed to separate the color with all its harmonic components from the luminance. Such "comb filters" produce a picture with considerably less chrominance-to-luminance distortion than normal filter techniques to generate a sharper displayed image.

As shown in Figure 1.70, the filtered-out chrominance is then decoded back into the original red, green, and blue video signals. Then, as shown in

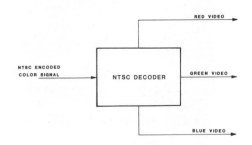

Figure 1.70
Decoder Separates the Encoded Signal
into Component Parts

Figure 1.71, the red signal is sent to control an electron beam that strikes only phosphors that glow red when energized. The green signal controls an electron beam that strikes only green glowing phosphors. The blue signal controls an electron beam that strikes only blue phosphors. A close examination of any color video monitor reveals separate red, green, and blue phosphors glowing even when displaying a black-and-white picture.

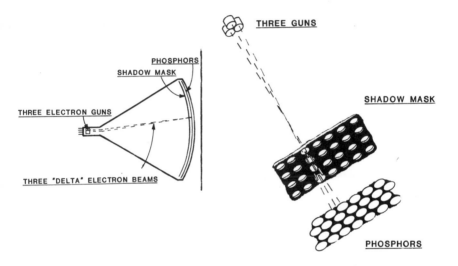

Figure 1.71
Delta Gun Picture Tube

The picture tube shown in Figure 1.72 uses a single gun to produce the three electron beams. The three beams are all in one horizontal plane and shine through an "aperture grill" of vertical slits to land on phosphors that glow red, green, or blue. These phosphors are applied in vertical stripes inside the front of the tube. The aperture grill stops the electron beam from reaching the phosphor deposits except when the proper phosphor color can be seen by the proper beam. The intensity of each of the three electron beams is independently varied before going through "electron optics" to be focused on the correct phosphor color.

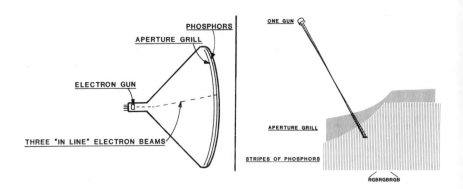

Figure 1.72
Vertical Stripe Picture Tube

Vertical stripe tubes (Trinitron®, Linytron®, Slot Matrix®, etc.) tend to give a sharp picture by masking out some of the noise and distortion in the signal and making ordinary vertical stripe tubes the choice for consumers and other end users. "High resolution" vertical stripe tubes with a greater number of color phosphor stripes are able to display small amounts of signal distortion (that may show up on the screen as smearing or outlining of sharp images) and noise (that shows up as "snow"). Such high resolution tubes are often used for critical testing and evaluation of the video signal in a television facility. If the picture looks clean on a critical high resolution display, it looks good on any normal display.

COLOR VIDEO PROJECTORS

Because of size and weight of limitations imposed by picture tubes in conventional color video monitors, the size of the viewing screen is severely limited. To allow television viewing by large groups, techniques have been developed to allow large screen projection of the color video signal.

As shown in Figure 1.73, the simplest video projection system uses a single glass lens to focus the image from a conventional picture tube on a screen. This approach is low in cost but presents significant technical limitations on the size and quality of the projected image. The image from the single lens projector often has color fringing because various colors are not refracted equally as they go through the single projection lens. The image is usually inverted, making it necessary to reverse the scan on the picture tube. Ordinary picture tubes are not designed to generate enough light output for projection use, for if a picture tube which normally provides an acceptably bright one square foot picture is projected to give a two square foot image, the light intensity of each picture detail drops by half because the same amount of light is spread over a larger area. This fixed relationship between picture tube output and image size limits the image size to a small magnification over the original size of the picture.

To overcome some of the size limitations, special high intensity picture tubes have been developed to increase the light output of the single lens projector. These tubes usually compromise the optimum operating relationships among image intensity, size of phosphor deposits, X-radiation, and the life of the phosphors of the high-intensity tubes.

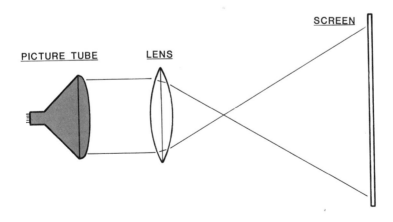

SCREEN

PICTURE TUBE   LENS

Figure 1.73
Basic Video Projector

SCHMIDT OPTICS

To overcome some of the limitations of the high intensity tubes, optical brightness intensification is often incorporated into the projection system. The majority of the better video projectors use a "Schmidt optics system" similar to what is shown in Figure 1.74, where independent super-bright picture tubes for variations of red, green, and blue are designed to shine the picture into parabolic mirrors before being projected onto the screen. The parabolic mirror concentrates the light through a correction lens to project a bright, undistorted image. Usually, the red, green, and blue video signals are fed to separate high intensity picture tubes with individual optics and appropriate phosphors to allow the image to be focused and "converged" (overlayed) for the best display.

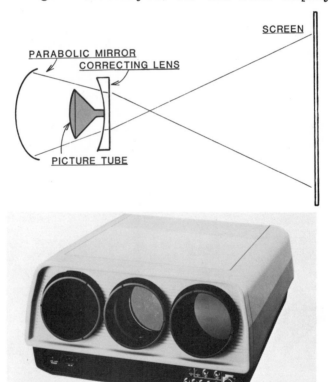

Courtesy of Sony Corporation

Figure 1.74
Schmidt Projection Optics System

LIGHT VALVES

But even with the Schmidt optics system and high intensity picture tubes, the size and brightness limitations of the system become evident once the picture is larger than about four feet (diagonal measure). To reliably generate larger images, totally different techniques are usually found.

The primary limiting factor in projecting a large television image is the maximum brightness output from the picture tube(s). If the light source used to produce the projected television image were changed from a glowing phosphor to a high intensity light bulb, the brightness of the image can be increased significantly. In such systems, the light for the projected image can be electronically controlled by a "light valve."

One of two current light valve projection techniques uses the light valve to vary the intensity of the light as it passes through the valve ("transmissive system"). The second technique varies the amount of light reflected out of the light valve ("reflective system").

The transmissive light valve system shown in Figure 1.75 uses the "diffraction" properties of light (ability of light to bend around obstacles in its path) to vary the projected light intensity. When an obstacle creates a shadow, some light bends around the obstacle to dull the focus of the shadow. The obstacle in the light valve is a coating of oil, the thickness of which is controlled by the strength of an electron beam. Some light is allowed to pass through while the remainder is reflected back or diffracted to pass through an opening in an output grating. To ensure that uncontrolled light directly from the source cannot pass through and only the diffracted light is able to pass through), the output grating is offset from an input grating so that no light can pass straight through the valve.

Figure 1.76 shows another technique for video projection where a "reflective light valve" is used to produce a bright image. A high intensity light shines on a mirror on which a layer of transparent oil has been coated. The thickness of the oil layer determines the angle of reflected light and whether the light is

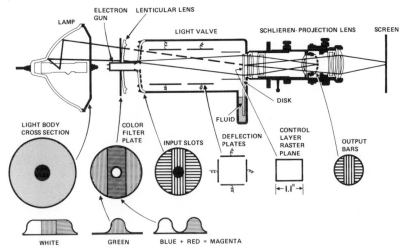

Figure 1.75
Transmissive Light Valve Projector

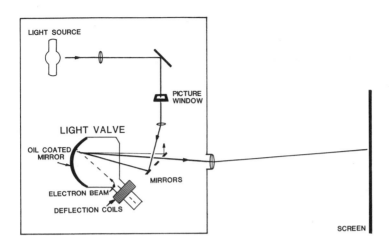

LIGHT SOURCE

PICTURE WINDOW

LIGHT VALVE

OIL COATED MIRROR

ELECTRON BEAM

DEFLECTION COILS

MIRRORS

SCREEN

Figure 1.76
Reflective Light Valve Projector

reflected back into the light source or allowed to pass on to the screen surface. An electron beam with an intensity that is varied by the video signal scans the oiled surface to vary the thickness of the oil layer. The light reflected from any given point in the area scanned by the electron beam is proportional to the strength of the electron beam at the instant the oil over that point is deformed. Figure 1.76 diagrams one-third of a color projector where individual light valves are used to modulate red, green, and blue light sources.

In applications requiring large color television displays in bright sunlight, several manufacturers have developed arrays of incandescent light bulbs, cathode ray tubes (CRTs), or liquid crystal displays (LCDs). These display systems sport brand names like Diamond Vision® and JumboTron®. The light bulbs, CRTs, or LCDs are commanded by a computer to glow a particular color, depending on the video signal fed to the computer. In essence, the bulbs and CRTs form the smallest definable picture element in the display. The resolution is relatively low, but the size may be large and the picture bright.

With all the technical sophistication built into the current video projectors, they still approach the limitations of the television system with its objectional horizontal line size (from scanning only 525 discrete horizontal paths), poorly defined picture details (caused by limiting the upper frequency of the encoded television video signal), and color rendition problems associated from operating the projection picture tubes at their maximum brightness limit. Many techniques continue to be developed to minimize the effects imposed by these specified signal limits.

COLOR MONITOR SET-UP

There are several controls on video monitors and projectors that must be properly adjusted to give an accurate display of the video signal. An established set-up procedure should be followed to ensure accurate and consistent display of the video signals.

On a color monitor or projector, after insuring that all operator controls for automatic circuits (AFT, AFC, ACC, etc.) have been turned OFF, a good black-and-white display needs to be obtained by turning down the COLOR or CHROMA control. On some monitors and projectors, the CHROMA or COLOR control may not have enough range to eliminate all the color in the picture when viewing a color video signal. In such cases, a black-and-white video signal should be used.

If the displayed picture is not a true black-and-white picture (and still has some color contamination), a qualified technician should be called to adjust the BIAS and SCREEN or GAIN controls. The technician has red, green, and blue BIAS controls to adjust the color purity of the dark picture details. Red, green, and blue SCREEN or GAIN controls are available for technician adjustment of the color purity of the displayed white picture details. In essence, the BIAS and GAIN adjustments establish the color balance of the displayed picture. Proceeding with the adjustment procedure without a good black-and-white display base results in a display set-up that can vary in quality from scene-to-scene.

After a true black-and-white picture has been obtained, the BRIGHTNESS control is adjusted to show details in black picture areas. The BRIGHTNESS control in a monitor internally adjusts the video signal in the same manner as the master PEDESTAL control on a camera. The CONTRAST is then adjusted to show details in the white picture areas. The CONTRAST control in a monitor internally adjusts the video signal in the same manner as the GAIN or IRIS control on a camera. BRIGHTNESS and CONTRAST should not be changed under constant viewing conditions, but adjustments may be necessary when the ambient light in the viewing area changes. BRIGHTNESS and CONTRAST should not be changed to accommodate changes in picture from scene-to-scene or from source-to-source. With a video monitor properly adjusted, approximately twenty different shades of gray (from black-to-white) can be readily discerned on the screen. (A black-and-white monitor is adjusted using the same procedure, except that there are no color circuits with which to be concerned.)

Adjusting a color video monitor or projector for a good black-and-white picture (generated by the luminance signal) provides a good picture "base" on which color can be added. Viewing a color signal, the COLOR or CHROMA control is then used to increase color saturation in the displayed picture to a desirable level. Many people watch "flesh tones" or the red color bar to watch for overdriving of the color and increased noise in the red picture areas.

The last step in setting up a color monitor or projector adjusts the TINT, HUE, or PHASE control for proper color rendition. The yellow or magenta color bars and flesh tones are the best indicators of hue. If an NTSC standard color bar pattern with -I, White, Q area is available (conforming to EIA Recommended Standard RS-189), the phase may be adjusted by equalizing the brightness of the -I and Q areas.

If the color video monitor or projector can be adjusted to display only the blue video signal (by turning off the red and green guns) and EIA (Electronic Industries Association) or SMPTE (Society of Motion Picture and Television Engineers) color bars are available, there is an alternative method of setting hue. (SMPTE has produced a video tape detailing proper set-up of a color picture monitor when using their standard bars.) When displaying color bars in the "blue only" mode, alternating blue and black vertical bars appear across the screen (as was discussed in the color bar primary color composition). By adjusting the HUE control for equal brightness of each of the black bars and each of the blue bars, the hue and the saturation of the display can be accurately adjusted. (The brightness, contrast, and saturation settings may also have some effect on the "blue only" display, but they should have already been adjusted for a pleasing monochrome picture.)

PURITY AND CONVERGENCE ADJUSTMENTS

Two last adjustments to be performed, if necessary, by a competent technician are purity and convergence. Convergence adjusts the path of the electron beams so that each beam energizes only the proper color phosphors. The visible results of bad convergence are color fringing and limited picture resolution. Convergence misadjustment may be easily confused with registration error in a camera.

When adjusting convergence, the technician uses an electronically generated pattern of white dots or a white grid on a black background, adjusting to eliminate any color fringing around the white areas. Since there are sometimes twenty or more interacting electrical and mechanical adjustments for convergence, the procedure should only be performed by a competent technician with plenty of time and the proper test equipment.

Purity refers to the evenness of color across the screen and is limited by the ability of the electron beam to fall only on the proper color phosphor at the proper time. A check of purity can be performed by displaying a full picture of a single color (including white) and looking for color changes anywhere in the display. Purity problems are caused by variations in the magnetic field around a monitor picture tube or by misalignment of the deflection coil yoke.

Degaussing the tube (moving the tube through a magnetic field to reduce the effects of any residual magnetic fields) frequently corrects an impure display. Many monitors have automatic degaussing by momentarily energizing a permanently installed coil of wire that surrounds the tube.

Any internal adjustment of the magnetic field around the picture tube needs to be performed by a technician. To be sure that an impure display is a fault from within the monitor, the surrounding areas should be cleared of magnets (on speakers, motors, etc.) that are capable of inducing the problem.

The only repeatable and accurate method of setting up a color picture monitor is to use a photometer and match the brightness for red, green, and blue in the dark and the bright picture areas. Some monitors like the one shown in Figure 1.76 have a built-in photometer that is placed against the screen and automatic circuits are commanded to adjust the various display parameters of the monitor.

Courtesy of Sony Corporation

COLOR PICTURE MONITOR BLOCK DIAGRAM

For reference, the block diagram for a typical color picture monitor is shown in Figure 1.78.

95

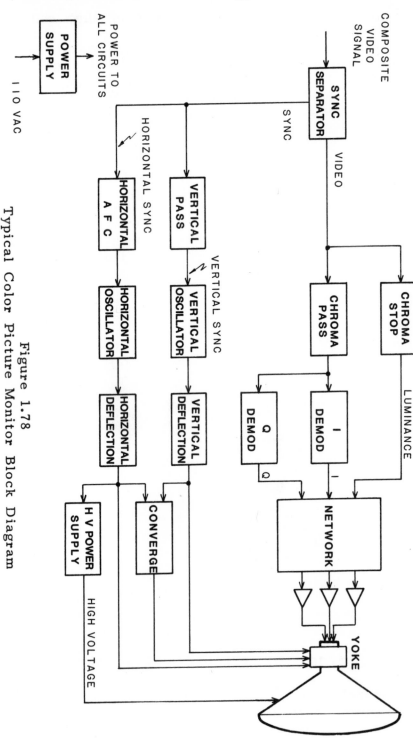

Figure 1.78
Typical Color Picture Monitor Block Diagram

QUESTIONS

1. If a lens has distortions that are most pronounced near the edges of the focused image, where in the television frame will these effects be apparent?

2. By what factor will the amount of light passing through a lens change when resetting the iris from f22 to f1.8?

3. Will a vidicon, Plumbicon®, or Saticon® produce the shallowest depth of field under the same scene illumination? Why?

4. What would be the picture effects if GAIN were boosted to compensate for use of a neutral density filter?

5. What procedure should be used to white balance a camera for aerial shots?

6. Which video signal parameters change between a green color bar and a magenta color bar?

7. If a thin object is horizontally centered in the frame, approximately how much time elapsed between the beginning of the scan and the object? How much time to the end of the scan? (Assume equal blanking times at the beginning and end of the scan.)

8. What are the height and width of the display of a 19" picture monitor?

9. Would a back focus adjustment of one pick-up tube affect registration adjustment of a multiple-tube color camera?

10. Describe a procedure to determine if a picture distortion is camera misregistration or a picture out of convergence.

11. Describe picture effects of too much image enhancement.

CHAPTER 2

THE
BASICS
OF
AUDIO

THE BASICS OF AUDIO

Sound is our interpretation of rapid variations in the pressure of air. For example, when hands are clapped together, air is rapidly squeezed from between the hands to form a wave of increased pressure that radiates from the hands. When this wave is detected by the ear, the brain interprets it as sound. Voice and music are waves of air pressure that have repeating patterns.

MICROPHONES

In the electronic audio system, a microphone is used to transform sound waves into an electrical description of those sound waves. Although there are many ways to construct a microphone, two types are currently in vogue to accomplish these transformations - dynamic and condenser.

In dynamic microphones, a sound-collecting diaphragm is attached to a coil of wire. As shown in Figures 2.1 and 2.2, this coil of wire moves through the magnetic field created by a permanent magnet, acting like a miniature electric generator. (Whenever a wire moves through a magnetic field, an electrical signal is generated.) When the wire moves in response to a rapidly increasing air pressure variation that has been "picked up" by the diaphragm, the "positive" motion of the magnet induces a "positive" electrical voltage in the coil of wire. A rapidly decreasing air pressure induces a "negative" electrical voltage. Thus, the output of the microphone is an electrical signal that varies its positive and negative voltage in proportion to the time (frequency) and strength (volume) of the received sound pressure waves.

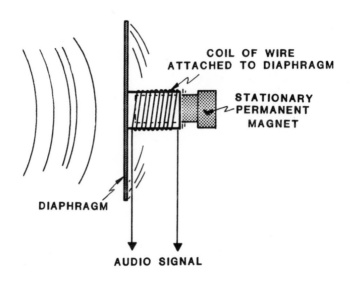

Figure 2.1
Typical Dynamic Microphone

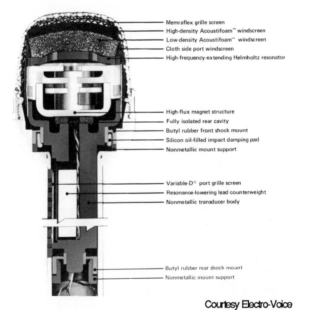

Memraflex grille screen
High-density Acoustifoam™ windscreen
Low-density Acoustifoam™ windscreen
Cloth side port windscreen
High-frequency-extending Helmholtz resonator

High-flux magnet structure
Fully isolated rear cavity
Butyl rubber front shock mount
Silicon oil-filled impact damping pad
Nonmetallic mount support

Variable-D® port grille screen
Resonance-lowering lead counterweight
Nonmetallic transducer body

Butyl rubber rear shock mount
Nonmetallic mount support

Courtesy Electro-Voice

Figure 2.2
Cutaway View of a Dynamic Microphone

In condenser or electret microphones, two parallel plates of conductive material serve as electric charge-holders and are separated by a nonconductive material (like air). If the distance between the plates varies, the ability of the plates to maintain a charge varies (closer together = less charge). If one plate is connected to a sound-collecting diaphragm and the other fixed as shown in Figure 2.3, the distance between the plates, and the electric charge between them, varies with the detected sound pressure waves. The varying electrical charge produced by inducing plate motion with sound waves is an electrical representation of the original sound information. To create the charge on the plates, a battery or AC power supply is required for proper operation. To detect only the *changes* in the charge caused by the sound waves, a transformer is used in the audio output circuit to block the DC voltage of the power supply.

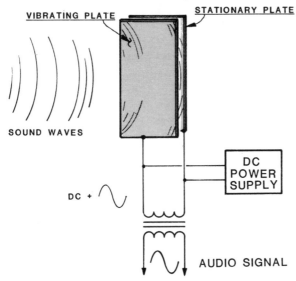

Figure 2.3
Typical Condenser Microphone

Microphones are also designed with a wide variety of mounting arrangements to ensure optimum usage qualities. From lavalier (around-the-neck), to hand-held, to boom mount, the selection depends on the intended application.

103

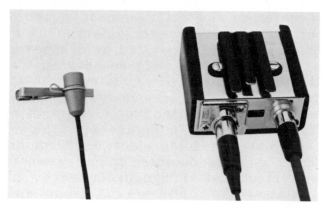

Courtesy Electro-Voice

Courtesy Panasonic Broadcast Systems

Figure 2.4
Typical Microphone Mountings

PICK-UP PATTERNS

To optimize the sound pick-up under various operating conditions, microphones are designed with various sound pick-up patterns. Figure 2.5 represents a top view of equal sound response where one can choose from omnidirectional (with sound pick-up in any direction about the microphone), cardioid (with a "heart shaped" pick-up), or unidirectional (with pick-up from only one direction). Several other special pick-up patterns are available from several manufacturers.

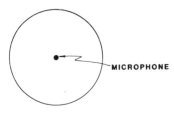

a) Omnidirectional

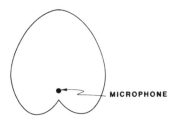

b) Cardioid

c) Unidirectional

Figure 2.5
Microphone Pick-up Patterns
(Top View)

To understand the different options in selection of a microphone pick-up pattern, Figure 2.6 poses the problem of trying to pick-up the sound from a string quartet seated around three microphones: one omnidirectional, one cardioid, and one unidirectional. The omnidirectional microphone picks up the sound evenly from all four musicians. The cardioid microphone picks up only three of the musicians at a time, and the unidirectional microphone only one musician. Sound *can* be heard from the musicians outside a given pick-up pattern, but their volume rapidly diminishes as they become farther outside the drawn pattern.

a) Omnidirectional

b) Cardioid

c) Unidirectional

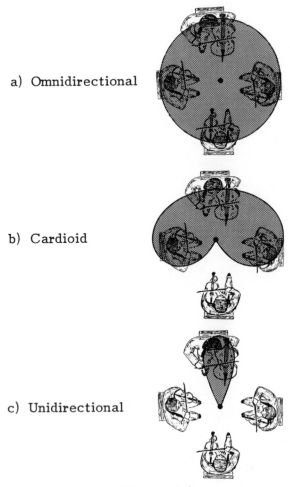

Figure 2.6
Pick-up Patterns Over a String Quartet

Specifications of pick-up patterns similar to that shown in Figure 2.7 show a line that traces the distance from the microphone to a sound source of constant power output where equal audio signal power is generated as the source moves around the microphone. If the constant sound source is located outside the line of equal response, it generates a lower power audio signal. The balance of several sound sources depends on their distance and relative positioning to the single microphone. Any adjustment of electrical power after a microphone generates the audio signal affects *all* source signal powers equally. Once the microphone generates the audio signal in response to several sound sources, the relationship among the signal powers (volume) of the various sources is fixed.

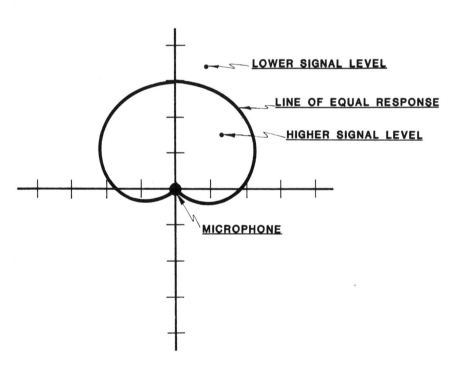

Figure 2.7
Typical Microphone Pattern

The same pick-up pattern characteristics can be used to eliminate unwanted sounds in the audio. When trying to pick out sound from talent surrounded by noisy machinery or air conditioning, judicious selection of a microphone with the most appropriate pick-up pattern can make the difference between acceptable output sound quality and unintelligible sound.

Microphones, by their construction, are susceptible to directly-coupled mechanical vibrations or to sound pressure variations caused by wind. Various methods of mechanical isolation (foam, springs, etc.) are often used to minimize microphone susceptibility to such mechanically induced noises. Besides outwardly visible mechanical isolation (like the boom microphone shown in Figure 2.4), some microphones with construction similar to that shown in Figure 2.8 offer internal shock-dampening designs to effectively isolate the microphone element from the outer case.

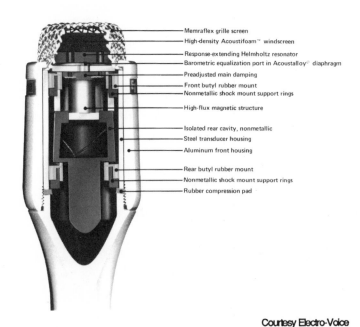

Courtesy Electro-Voice

Figure 2.8
Cutaway View of a Mechanically Isolated Microphone

108

Moving the microphone closer to the sound source reduces the effects of unwanted sounds from vibration or noise sources by effectively making the desired sound louder than the undesired sounds. This effect is the result of the spreading of the sound energy to fewer and fewer air molecules as the distance from the source is increased. The "presence" (perceived proximity of the sound source) and signal quality are also enhanced because there are fewer air molecules between microphone and sound source to dissipate the energy from high-frequency sound pressure waves.

AUDIO MIXERS

It is frequently desirable to combine signals from several microphones (or other sources) into a single "program audio" signal. To eliminate distortion and provide operator control over the relative volume of each audio signal source, an electronic "mixer" similar to that shown in Figure 2.10 is used.

Figure 2.9
Typical Audio Mixer

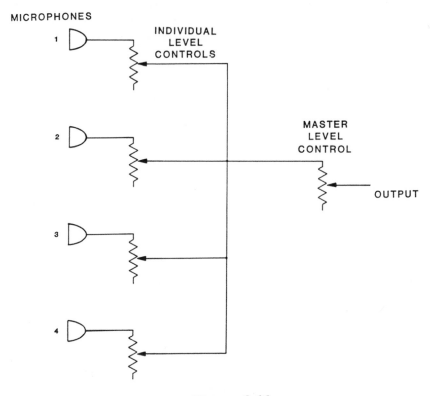

Figure 2.10
Simplified Audio Mixer Schematic

Each audio signal source that is input to the mixer is given an individual level control. Once each of the signal sources is balanced to form the desired audio scene, the overall level of the finished product can be controlled by a master level control. The master level control is used to adjust the mixer to optimize the output signal power.

On audio mixers designed for stereophonic operation, controls are found to place each individual sound source at a particular location within the audio scene. Such "panning" controls adjust the relative audio signal strength of a given source between the left and the right channels.

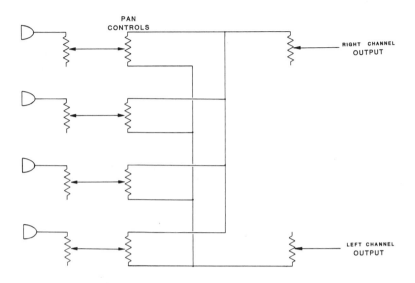

Figure 2.11
Simplified Stereophonic Mixer Schematic

Some mixers also provide "reverberation" controls to produce an "echo" effect. As shown in Figure 2.12, the circuits that are used in reverberation techniques simply take a small amount of the output signal, delay it, and reinsert the delayed signal back into the output circuitry. Controls for the adjustment of amount of delay and the power of the delayed signal (relative to the undelayed signal) is usually provided.

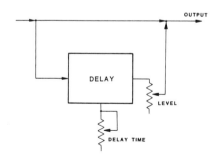

Figure 2.12
Typical Reverberation Circuits

111

AUDIO EQUALIZERS

Once the audio signal is formed, there is a limited amount of change that can be performed. Some signal frequencies can be isolated and then increased or decreased in power, leaving the other frequencies alone. For example, the high frequency signal components of a signal that represent the treble tones in the original sound can be "boosted" so the recreated sounds seem "crisper." Other examples of power transfer versus frequency are shown in Figure 2.13. The alteration of the strengths of the various frequency components of an audio signal are performed by "equalization" controls.

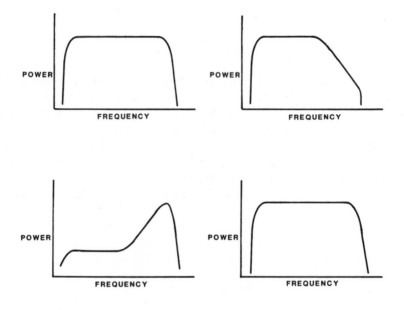

Figure 2.13
Frequency Response Graphs of an Equalized Signal

LOUDSPEAKERS

To turn the audio signal back into sound, the signal is amplified (increased in power) and fed to a loudspeaker. The permanent magnet speaker, by far the most common type of speaker, looks similar to a dynamic microphone and is shown in Figure 2.14. The theory used in dynamic microphones and permanent magnet speakers is so similar that speakers are used as microphones in noncritical applications. For a speaker, the magnet is not moved through the coil of wire in response to sound waves as in a microphone; instead, the magnetic field generated by the coil of wire is varied by the audio signal. This varying magnetic field then attracts or repels the permanent magnet in response to the audio signal (remembering that like magnetic poles repel and opposite magnetic poles attract). The moving coil is attached to a flexible paper or plastic cone to physically move huge quantities of air molecules when creating sound waves.

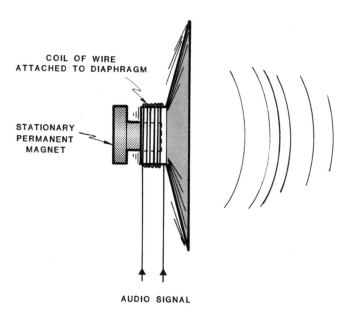

Figure 2.14
Typical Permanent Magnet Loudspeaker

INTERCONNECTING AUDIO EQUIPMENT

Audio equipment input and output designs have not been standardized, unlike most video equipment, and must be matched to maintain a high quality signal through a complex system. In general, there are three variables that should be matched to optimize signal transfer between any two pieces of equipment:

IMPEDANCE

SIGNAL LEVEL

BALANCE CONFIGURATION

IMPEDANCE

Impedance (abbreviated "Z") is opposition to the flow of electricity. It is similar to, but not exactly the same as "resistance."* If the output impedance (Z_{out}) of one piece of equipment does not match the input impedance (Z_{in}) of another piece of equipment, direct connection between the two units could introduce significant distortion.

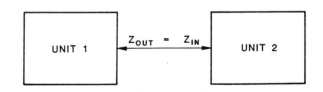

Figure 2.15
Matching Impedance

---

*The term "Resistance" refers to opposition to the flow of a constant voltage (DC) signal. Resistance is what a common multimeter measures. The term "Impedance" refers to opposition to the flow of varying voltage (AC) signals like audio and video. Impedance consists of resistance to the effective DC component of an AC signal plus "reactance" to the effective AC component of the signal. The value of the reactance component (and the total impedance) depends on the frequency of the applied signal. Impedance also describes the phase response of a circuit.

To understand the importance of impedance, one must consider some of the characteristics of a signal as it travels from one circuit, down a transmission line, and into the next circuit. An analogy may be drawn to water pipes where the water current flows smoothly until it comes to a pipe joint. If the sizes of the pipes joined together are of equal size, the water flows through smoothly. If the water flows from a large pipe to a small pipe, some of the water is reflected back at the joint, creating ripple distortion, and interfering with the forward flow. If the effects of impedance on the flow of electrons is similar to the effects of pipe size on water flow, one conclusion may be drawn: electrical signal transfer with a minimum of loss and distortion occurs when the two circuits involved have equal impedance.

As shown in Figure 2.16, when an impedance mismatch occurs, some of the signal is bounced back from the point of impedance mismatch to interfere with the original signal. If the reflected signal is 180° out of phase with the original signal, the power of the signal is reduced but no distortion is produced. If the reflected signal is out of phase with the original signal, both power reduction and distortion occurs.

Matching between two different design impedances can be accomplished by using a "pad" network made of resistors or a specially designed audio matching transformer as shown in Figure 2.17. These pads and transformers are designed to let the equipment output see its characteristic impedance and the equipment input see its impedance. For information on pad design, see Appendix A.

A special condition exists when the output impedance of one device is low compared to the input impedance of the equipment connected to it. As shown in Figure 2.18, a 1:10 ratio of $Z_{out}:Z_{in}$ allows the input to "bridge" the output. Here, one resistor of the value of the output impedance can be placed across the line and one resistor of the value of the input impedance inserted in the line between the terminating resistor and destination provides the best match between the two pieces of equipment.

ORIGINAL SIGNAL   REFLECTED SIGNAL

LOAD

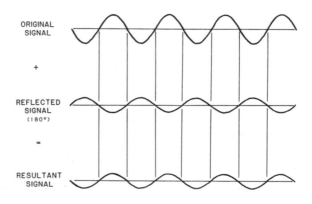

ORIGINAL
SIGNAL

+

REFLECTED
SIGNAL
(180°)

=

RESULTANT
SIGNAL

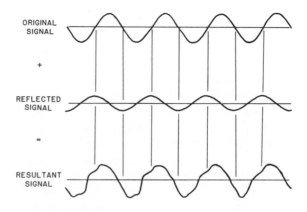

ORIGINAL
SIGNAL

+

REFLECTED
SIGNAL

=

RESULTANT
SIGNAL

Figure 2.16
Effects of Impedance Matching

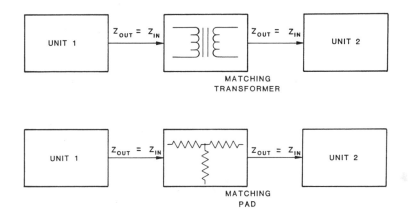

Figure 2.17
Impedance Matching Techniques

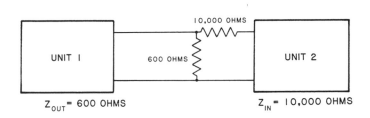

Figure 2.18
Impedance Matching of a Bridging Input

The intent of the design of a bridging input is to allow direct connection of several inputs to one output; however, there is a limit to the number of inputs connected because the total impedance produced by interconnecting several bridging inputs *decreases* according to the following formula:

$$\frac{1}{Z_{Total}} = \frac{1}{Z1_{in}} + \cdots + \frac{1}{Zn_{in}}$$

Two 10,000 Ohm inputs bridged together creates an apparent 5,000 Ohms impedance. For three 10,000 Ohm bridging inputs connected together, the apparent impedance drops to 3,333 Ohms.

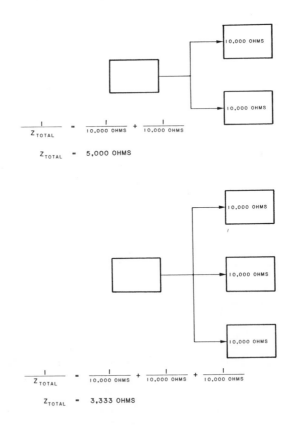

Figure 2.19
Impedance of Bridging Inputs

It is not unusual for one corrected impedance mismatch to make the difference between bad system sound and good sound output. But even if one impedance mismatch does not produce noticeable sound degradation, many mismatched connections in a system can significantly reduce the quality of the output sound. The distortion products generated in one mismatch create even more distortion products in the next mismatch. The next mismatch creates distortion products from the distortion products. A properly matched system always has a lower *apparent* signal level than an unmatched system, but the apparently higher signal in the unmatched system has considerably higher distortion content.

SIGNAL LEVEL

Signal levels between pieces of equipment should also be matched. Because electronic circuits must be designed to accommodate a limited range of signal levels, the output circuitry of one piece of equipment should provide the optimum signal level to the input circuitry of the next piece of equipment in the system.

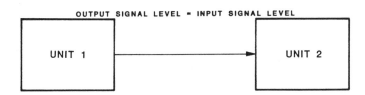

Figure 2.20
Matching Level

For ease in describing and calculating signal levels in a complex system, most audio signals are expressed as the logarithm of a ratio of power, voltage, or current between two signals. The unit used in describing the electrical ratio is the "Bel," or the more common "deciBel," abbreviated "dB." The dB is related to signal *power* by the formula:

119

Power Ratio (in dB) = 10 X LOG $\dfrac{\text{Power of Signal A}}{\text{Power of Signal B}}$

Where

$$\frac{\text{Power of Signal A}}{\text{Power of Signal B}}$$

indicates the gain (if greater than 1) or loss (if less than 1) of the signal.

The log of a number is the value to which the base 10 must be raised to equal the number. For example:

$$\log 1000 = \log 10^3 = 3$$

$$\log 100 = \log 10^2 = 2$$

$$\log 1/100 = \log 10^{-2} = -2$$

$$\log 16 = \log 10^{1.2} = 1.2$$

$$\log 1/16 = \log 10^{-1.2} = -1.2$$

$$\log 2 = \log 10^{0.3} = 0.3$$

$$\log \tfrac{1}{2} = \log 10^{-0.3} = -0.3$$

Using this understanding of logarithms, power ratios can be readily expressed in dB:

Power of Signal A = 2 Watts

Power of Signal B = 1 Watt

dB Ratio = 10 log (2/1) = 3 dB

Power of Signal A = 4 Watts

Power of Signal B = 2 Watts

dB Ratio = 10 log (4/2) = 3 dB

Power of Signal A = 16 Watts

Power of Signal B = 1 Watt

dB Ratio = 10 log (16/1)   = 12 dB

Power of Signal A = 1 Watt

Power of Signal B = 16 Watts

dB Ratio = 10 log (1/16)   = -12 dB

As shown in Figure 2.21, Signal A is normally the output from a piece of equipment or system and Signal B is the input. This means that equipment or systems with signal *power* gain or amplification has a positive value of dB Ratio and equipment that has a signal power loss has a negative value. This ratio is valid for power-based systems only; voltage or current-based systems is reflected by different numbers in dB-expressed relationships. For a more complete discussion of logarithmic relationships, consult an algebra text.

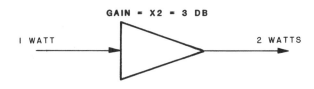

GAIN = X2 = 3 DB

I WATT   2 WATTS

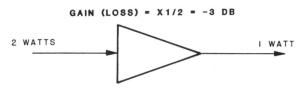

GAIN (LOSS) = X 1/2 = -3 DB

2 WATTS   I WATT

Figure 2.21
dB Expressions of Gain and Loss

Mastering Television Technology

To help standardize power specifications, signals are frequently compared to a standard reference power. The most often used standard in audio is the "dBm," where the reference is 1 milliWatt (0.001 Watt) of power into a 600 Ohm impedance.

$$\text{Power Ratio (in dBm)} = 10 \times LOG \frac{\text{Signal A Power}}{0.001}$$

Using this formula,

$$0 \text{ dBm} = 10 \times \log \frac{.001}{.001} = 10 \times \log 1 = 10 \times 0 = 0$$

0 dBm = 1 milliWatt of signal into 600 Ohms

Any difference from the 0 dBm reference is a ratio of power.

$$+3 \text{ dBm} = 10 \times \log \frac{.002}{.001} = 10 \times \log 2$$

+3 dBm = 2 milliWatts (double the power)

$$-3 \text{ dBm} = 10 \times \log \frac{.0005}{.001} = 10 \times \log 1/2$$

-3 dBm = ½ milliWatt (half the power)

Note that 0 dBm at 600 Ohms impedance is not equal to 0 dBm at 150 Ohms impedance and correlation between the two requires use of conversion factors. Consult the audio reference texts in Appendix B for further information.

Frequently a special meter is used to measure audio power. This meter, calibrated in volume units ("VU"), indicates the average peak instantaneous power. Normally, the audio power is adjusted so the VU meter peaks at "0." The calibration of the VU meter is the desired optimum power for operation of that particular circuit and *may not correlate to any particular Audio standard.* Zero VU may or may not equal 0 dBm. For example, "0" on the VU meter of a tape recorder usually indicates the point of optimum signal drive for the recording processes.

Figure 2.22
Two VU Meters on a Typical Video Tape Recorder

In general, the specified optimum audio signal levels of interconnected equipment should be matched. A 0 dBm, 600 Ohm output should be connected to a 0 dBm, 600 Ohm input. If the output signal from a piece of equipment is too high, the level can usually be reduced by internal equipment adjustment. If there is no internal adjustment, or if the adjustment range is insufficient, resistive pads may be required to reduce the output signal level or match impedance. As shown in Figure 2.23 if the maximum signal level output from a piece of equipment is lower than the minimum usable input level of the connected equipment, an amplifier is required to increase the power of the signal to a usable level.

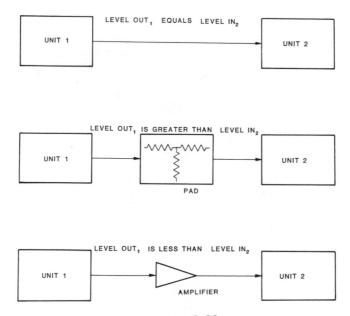

Figure 2.23
Level Matching Techniques

AUTOMATIC GAIN CIRCUITS AND LIMITER CIRCUITS

Special "automatic gain circuits" (AGC) are available on many types of equipment to adjust a wide range of input audio (and video) signal levels to the optimum recording level. As shown in Figure 2.24, the AGC examines the average voltage of the signal over a period of time. (The length of time is dependent on the design of the AGC.) If the incoming signal is low *on the average*, the AGC boosts the level before recording. If the incoming signal is high *on the average*, the AGC reduces the level.

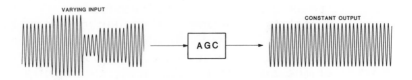

Figure 2.24
AGC Operation

124

Problems are encountered when there is a sharp transition between a low and a high signal level. Because of the built-in delay, the AGC's response to a fast low-to-high signal level change is frequently late. As shown in Figure 2.25, the circuit originally adjusted itself to provide a large power gain for the weak signal. After the stronger signal arrives, the AGC takes the time that was designed into the circuit to average the signal power and optimize the gain for the strong signal. A soft music passage produces an optimum listening volume through an AGC amplifier. Within limits, this volume is maintained regardless of the changes in volume of the original sound. Problems arise when the music suddenly changes from soft to loud. The AGC has raised the amplification of the signal to the optimum listening volume. When the loud passage starts, the AGC's gain is too high and the volume too loud. As the amplifier averages more and more of the loud signal with the soft signal, the gain of the AGC circuit is reduced until the loud passage is a comfortable volume. (The effects of an AGC on a video signal are identical, with substitution of "bright" for "loud.")

Exactly the reverse happens when a high-to-low signal level transition is detected by the AGC. The weak signal is not amplified to the optimum gain until the AGC senses the weak average signal level.

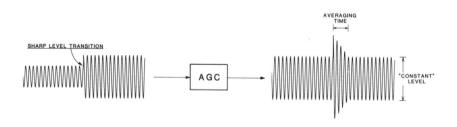

Figure 2.25
AGC Action on Sharp Signal Level Transitions

A limiter circuit is often used with an AGC, or by itself, to allow the operator to set the maximum attainable signal level. As shown in Figure 2.26, a limiter does not amplify weak signals. Just the power of strong signals that rise above a preset limit are reduced to an optimum level.

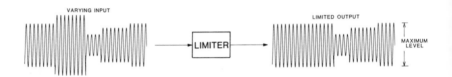

Figure 2.26
Limiter Operation

Different applications dictate whether an AGC, a limiter, or both should be used. With a single operator in portable production, the AGC provides flexibility for variations in microphone coverage or for signals with widely varying levels. In an application where music is being recorded, the limiter is often desired to allow soft passages to remain soft while proper equipment operation is preserved during loud passages.

In general, use of an AGC should be carefully examined because of the operational limitations. It is rare that the operational limitations outweigh the distortions that the circuit produces. A limiter circuit gives the best of both worlds.

BALANCE CONFIGURATION

The balance configuration of the signals between two pieces of equipment is the last parameter that must be matched for proper system interconnection.

As shown in Figure 2.27, an unbalanced signal flows entirely over only one wire (relative to a reference). The signal voltage is the voltage difference between the signal wire and a reference wire (that is frequently connected to equipment "ground").

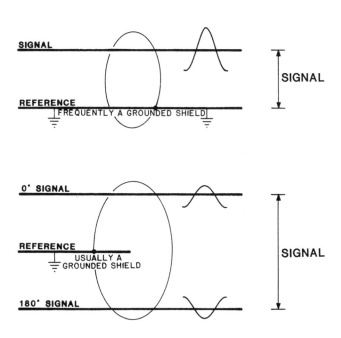

Figure 2.27
Balanced and Unbalanced Signal Configurations

A balanced signal is carried over two wires with voltages electrically isolated from the reference "ground" voltage. The instantaneous *difference between* the voltages of the signals on the two wires is the resultant audio signal. The signal voltage on one wire mirrors the other by being 180° out of phase. A balanced system is used where stray electrical interference may be picked up (such as power line "hum" or fluorescent lamp "buzz"). In a balanced system, the electrical interference is usually picked up equally on both signal-carrying wires, changing the voltage on both wires equally, but not changing the voltage difference between the two wires. Since a balanced system is concerned with voltage difference between the two signal carrying conductors, effects of the interfering fields is minimized.

The ground reference in both balanced and unbalanced systems frequently creates a hum in the output signal when a "ground loop" is present. Ground loops generate an interference because of a difference in potential of the grounds of two interconnected pieces of equipment. This potential difference and the resultant hum can be reduced by powering the equipment from mains power of the same phase, keeping audio cables short, having an isolated path to ground from each piece of equipment, and disconnecting (or "lifting") balanced cable shields (at one end of a cable only). Notice that a ground loop may be formed by unbalanced video cables even though the balanced audio cables have been properly conditioned. In some extreme cases, isolation transformers or couplers are required on each cable.

Interconnection between balanced and unbalanced configurations must be accomplished as shown in Figure 2.28 through a transformer called a balun - "*bal*anced/*un*balanced." As shown in Figure 2.29, attempts to directly connect between balanced and unbalanced signals results in one-half of the signal being thrown away. Circuits designed to accept balanced signals expecttwice as much signal if half the signal is being thrown away. With less signal available, the necessary amplification will be accompanied by the noise always created in electronic circuits. This increase in noise may not be apparent in practice (and such a direct mismatch of signal types is frequently dictated by emergency), but the effects of improper balance matching is cumulative and should not be considered acceptable routine practice in audio systems where matching must repeatedly occur.

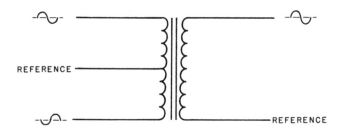

Figure 2.28
Balun Schematic

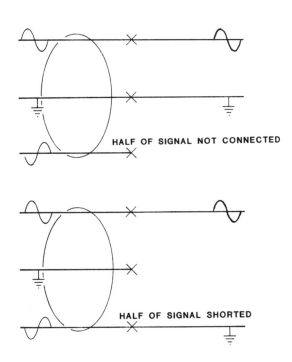

HALF OF SIGNAL NOT CONNECTED

HALF OF SIGNAL SHORTED

Figure 2.29
Improper Interconnection of
Balanced and Unbalanced Signals

LINE, MICROPHONE, AND SPEAKER SIGNALS

Audio equipment is typically designed to work with signals of three types of level - microphone, line, and speaker. Circuits designed to use the output from a microphone amplify the low power (usually about -55 dBm) that comes directly out of a microphone to a usable power that is less susceptible to outside electrical interference. A line-level circuit is designed to use signals that have already been amplified (to around 0 dBm) and can use that signal with little or no boost in level. Notice that a line-level signal can be connected to a microphone input, *if impedance, level, and balancing condition requirements are matched* with pads or transformers. A line-level input may accept a microphone-level signal if the signal is amplified to get the signal power to a usable level.

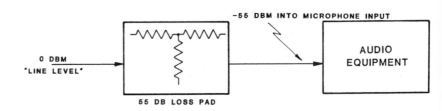

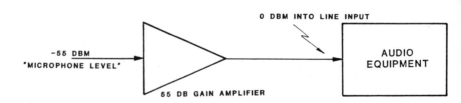

Figure 2.30
LINE and MICROPHONE Inputs

The output signals from speaker-level circuits have been greatly amplified by a "power amplifier." Most amplifiers are specified to have a peak power output in Watts or to have an average ("RMS" or "Root Mean Square") power output. Amplifiers are designed with a characteristic output impedance and balance configuration.

Impedance, Signal Level and Balance Configuration are important points to remember in audio systems. But, as we shall see in the section on video terminations, the three parameters are important when using any type of signal, including video.

QUESTIONS

1. What type of microphone pick-up pattern should be used in noisy field production locations?

2. When using two microphones to pick-up the same sound source, what precautions should be taken to maximize sound quality?

3. Is a balanced or unbalanced signal conveyed via two-conductor phone plugs? Is a balanced or unbalanced signal conveyed via three-conductor XLR connectors?

4. Can a "speaker output" be connected directly to a "line input?" How can a "speaker output" be connected to a "line input" to minimize distortion?

5. What is the term used to describe 60 Hz power line interference? What is the term used to describe interference from fluorescent lamp ballasts and other interfering sources?

6. What accessory is required for operation of a condenser-type microphone? Is this accessory required for operation of a dynamic-type microphone?

7. Would a condenser-type or dynamic-type microphone produce a truer rendition of high-frequency sounds? Why?

8. When designing a sound stage, why are extreme measures taken to reduce ambient noise?

9. What is the peak level in dBm of the output signal from an amplifier rated at 2 Watts, peak? How can this signal be used as the input to a circuit that has a rated input of 0 dBm?

10. How much amplification in dB has a microphone signal (average level of -55 dBm) undergone if it is output from an amplifier rated at 75 Watts RMS? How many TIMES greater is the speaker output level than the microphone input level?

CHAPTER 3

RECORDERS

Once sound and picture information are converted into audio and video signals, there is often a need for a way to store those electrical signals for later use. To date, the easiest medium to use in the storage of electrical signals is magnetic tape.

RECORD, PLAY AND ERASE MODES

As shown in Figure 3.1, magnetic tape is a strip of plastic with a thin coating of a material that readily responds to a magnetic field. This "oxide" coating is made up of tiny magnetic particles that are originally arranged with random magnetic orientation similar to that shown in Figure 3.2. During recording, the pattern of *magnetic orientation* of the particles in the coating is rearranged in response to the strength of an applied magnetic field. Notice that the particles do not physically move, but simply alter the positions of their individual "north" and "south" poles to form a particular pattern.

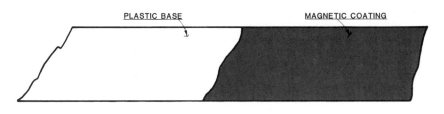

PLASTIC BASE   MAGNETIC COATING

Figure 3.1
Magnetic Tape

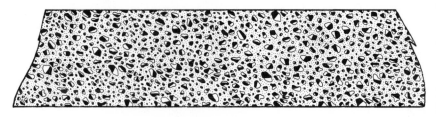

Figure 3.2
Magnetically Susceptible Particles
in the Oxide Coating of
Magnetic Tape

If the tape is moved through the magnetic field of an electromagnet, the strength of which is varied by an electrical signal as the tape passes by, the particles alter their magnetic orientation in response to the strength and polarity of the magnetic field that happened to be generated as they were passing by. As shown in Figure 3.3, even after the magnetic field has been removed, the particles retain their altered magnetic orientation. This pattern is an accurate record of the voltage strengths of the original signal.

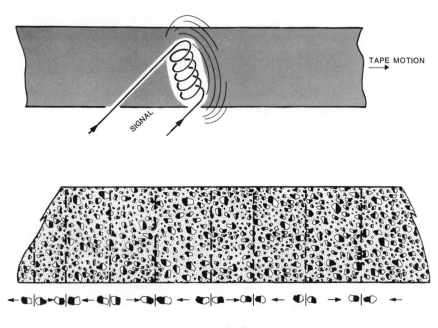

Figure 3.3
Magnetic Pattern in Tape after Recording

To record a true representation of the applied signal, enough magnetic particles must be exposed to the magnetic field fast enough to record all the rapid changes in the magnetic field (and the applied signal). As shown in Figure 3.4, if the particles are not moved through the field fast enough, a complete signal cycle may cause the magnetic field to expand and collapse while only one particle is exposed. The lone particle could not record any change in signal, retaining a magnetic orientation corresponding to the signal condition at the instant that the effects of the magnetic field were removed. When reading the pattern on the tape, it appears that no cycle of the recorded signal voltage ever occurred.

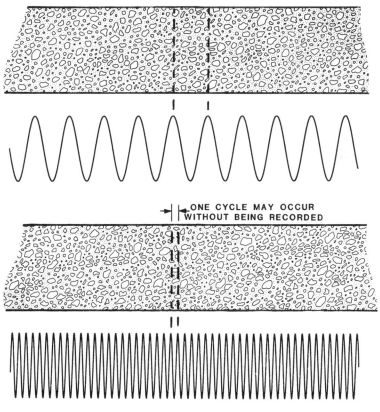

ONE CYCLE MAY OCCUR
WITHOUT BEING RECORDED

Figure 3.4
Missing a Cycle of Signal
by Exposing Too Few Particles

Detecting the magnetic pattern that may be in the oxide material on the tape uses the varying magnetic field to induce a voltage in a coil of wire. (When a magnetic field moves through a coil of wire, a voltage is induced in the coil.) If the tape is moved in the same direction and at the same speed as in the recording process the induced voltage is of the same rate and relative amount of voltage changes that were originally placed across the coil of wire in the record head. This induced voltage is amplified and becomes the playback signal. The same coil of wire is sometimes designed to both record and playback the signal.

The coils of wire used in tape recording and playback are constructed inside the magnetic "heads" of the recorder. The heads physically look like Figure 3.5 with the gap between two "poles" allowing generation or detection of a weak magnetic field. On many tape recorders, there are three types of heads:

RECORD - impresses the electrical signal voltage variations into the tape as magnetic variations.

PLAY - detects the magnetic variations on the tape and convert them into voltage variations of an electrical signal.

ERASE - removes any <u>pattern</u> of magnetic orientation of the <u>particles</u> in the tape.

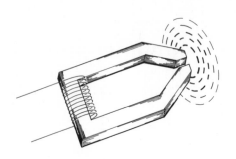

Figure 3.5
Magnetic Head Construction

The signal applied to the erase head is of constant voltage. With no *change* of magnetic field of the tape to induce a voltage in the play head, no signal is induced. Notice that once the erase head removes any previously recorded magnetic pattern from the tape, the original signal can no longer be retrieved.

THREE HEAD ARRANGEMENT

One particular arrangement of heads optimizes the magnetic tape recording processes. As shown in Figure 3.6, the first head encountered by the tape as it moves across the transport is the erase head. By first removing any previously recorded magnetic patterns, the exposure to the unchanging field of the erase head eliminates possible interference between any old magnetic pattern on the tape with the pattern that is to be recorded. Sometimes the brief exposure to the erase head is not enough to remove all the magnetic variations in the tape, and a "bulk eraser," similar to the one shown in Figure 3.7, is used before placing the tape on the recorder. The bulk eraser has a strong magnetic field to more completely erase the magnetic variations than the exposure to an erase head.

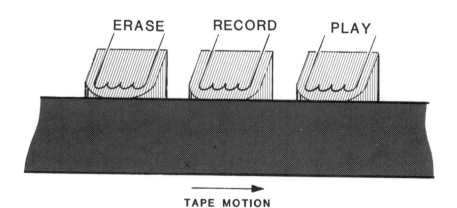

Figure 3.6
Typical Three-Head Arrangement

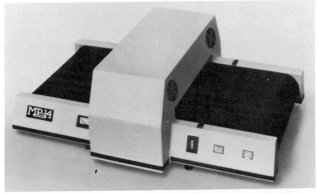

Courtesy Data Security, Inc.

Figure 3.7
Typical Bulk Eraser

The second head in the path of the tape is the record head. After the tape has been erased of all previously recorded magnetic pattern, a "fresh" tape is available to record the pattern produced by the new signal.

The third head plays back the information recorded on the tape. With this head arrangement (erase, record, then play), the recording can be played back to confirm recorded quality immediately after it is recorded. This "confidence head" arrangement for immediately playing back the signal is frequently seen on better audio recorders, but the cost of video heads is *usually* so much higher than audio heads, that a combination of the play and record functions into one head is frequently found.

There are two other important parts to a tape recorder that ensure a constant speed and tension of the tape past the heads - the "capstan" and the "pinch roller." The motor driven capstan rotates to pull the magnetic tape across the transport. The tape is squeezed between the metal capstan and a rubber pinch roller to reduce tape-to-capstan slippage and maintain constant tape tension and speed. The capstan and pinch roller are frequently assisted by motors that drive the feed and take-up reels. These "reel motors" help to maintain constant tension and speed of the tape as the quantity of the tape on the reels vary.

Constant tension and speed of the tape is essential for accurate playback of the recorded information. As shown in Figure 3.8, if the tape is faster in playback than in record, the magnetic variations are detected at a faster rate (frequency) than they were recorded. When the playback speed of an audio tape is faster than the record speed, the playback signal sounds higher in tone than the original sound source. An extreme example is the "chipmunk" sound heard when the tape is shuttled past the heads in the fast-forward mode.

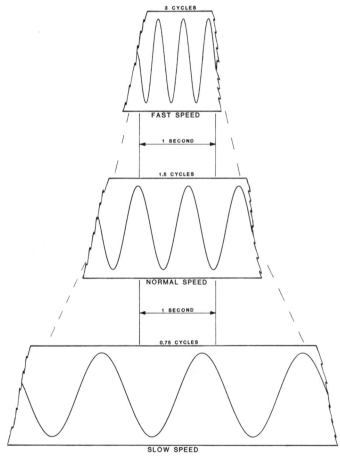

Figure 3.8
Differences in Tape Speed
During Record and Playback

Even with constant control of the tension and speed of the tape past the heads, variations in the played back signal can be caused by variations in the size of the tape itself. If a tape is stretched between the time of recording and the time of playback (by improperly adjusted tape transport tension or rough handling), the magnetic variations on the tape are pulled farther apart. To ensure that tapes are not stretched beyond normal limits, a tension gauge, head protrusion gauge, and spindle height gauge are frequently used in the routine maintenance of transports.

VIDEO RECORDING

Recording of the video signal is sometimes accomplished exactly as described in the previous sections, but the enormous amount of information that must be recorded to accurately describe a moving television picture usually dictates a faster speed of tape past the video record/playback head. With about 4 MILLION pieces of information to be recorded or played back every second (instead of the approximate 15 THOUSAND pieces in audio signals), special techniques are required to expose enough of the magnetic particles in the tape to accurately record and play back the video signal. Above the minimum number of magnetic particles it takes to record the most coarse of picture details in the video signal, the more magnetic particles that are exposed in a given length of time, the better the resulting picture resolution.

To increase the speed of the tape past the video heads could mean that thousands of feet of tape might be required to record a one-hour video program; but in the interest of economy, video tape recorders (VTRs) and videocassette recorders (VCRs) are usually designed to move *both* the tape and the heads. VTR's that use 1/2-inch or 3/4-inch wide tape usually have at least two video heads, similar to those shown in Figure 3.9, mounted 180° apart on a spinning cylinder. (Some VTRs have additional heads for slow motion, search, extended play, still play, or other special effects; yet, only two heads are used in the highest

144

quality mode.) Figure 3.10 shows how the tape slowly spirals around this spinning video head "drum" and across the transport. As shown in Figure 3.11, the two heads are electronically switched to ensure that the head that is in contact with the tape at the time is used to record or play back the signal. Such "helical scan" (from spiral or "helix") techniques allow a fast "writing speed" of the video information diagonally across a tape, while not consuming too much length of tape.

Figure 3.9
Typical Video Head

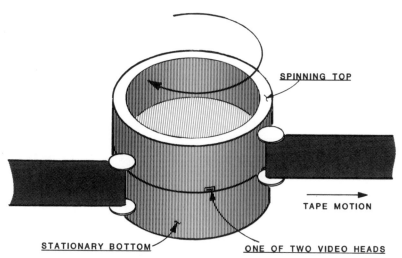

SPINNING TOP

TAPE MOTION

STATIONARY BOTTOM

ONE OF TWO VIDEO HEADS

Figure 3.10
Tape Wrap Around the Video Head Drum

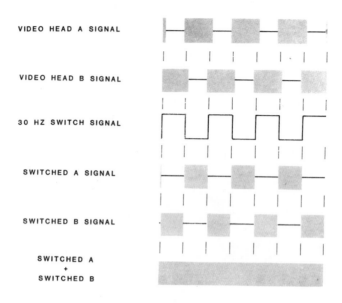

VIDEO HEAD A SIGNAL

VIDEO HEAD B SIGNAL

30 HZ SWITCH SIGNAL

SWITCHED A SIGNAL

SWITCHED B SIGNAL

SWITCHED A
+
SWITCHED B

Figure 3.11
Head Switching

Since the audio signal does not contain as many pieces of electrical information in a given time as the video, the audio signals are recorded on a "track" along the edges of the tape as in a standard audio tape recorder. To create this "longitudinal" audio track, a transport similar to the one in Figure 3.12 is designed so the tape is moved past stationary audio heads after it has moved past the spinning video heads.

All of the information must be recorded on the tape within tracks that adhere to strict standards for that tape's "format." The format dictates location of the information, the speeds at which the information is recorded, and the physical properties of the tape itself to allow "interchange" of a tape from one VTR to another of the same format. Figure 3.13 shows a photograph of a ½-inch VHS-format video tape that has been "developed." (Actually iron filings suspended in solution were applied to the tape and the solution allowed to evaporate.) Other common video tape formats are detailed in Appendix B.

146

Figure 3.12
Typical Video Cassette Recorder Transport

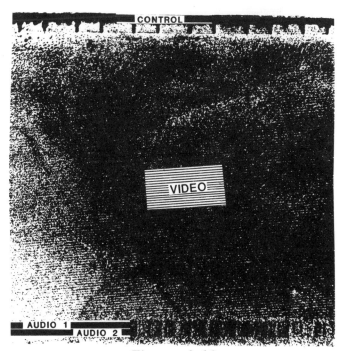

Figure 3.13
Photograph of
Developed VHS-Format Video Tape

SERVO CIRCUITS

Because of the huge amount of information that must be packaged in the small space of the tape used in a VTR, the speed and positioning of the tape as it moves across the transport must be precisely maintained. When the speed and positioning of the tape relative to the heads is not properly controlled, the signals as read off the tape has visible and audible distortion.

Special "servo circuits" are used to maintain close tolerance of the positions of video head paths across the tape and the speed of the tape across the transport. As shown in Figure 3.14, servo circuits create an electrical signal that is based on any *differences* in timing or amplitude between two signals from different sources. In a VTR, the signal that a servo circuit generates to describe the difference between two signals is used to command a motor to speed up or slow down for correction of any deviation from normal operation (as described by the differences between the original reference signals).

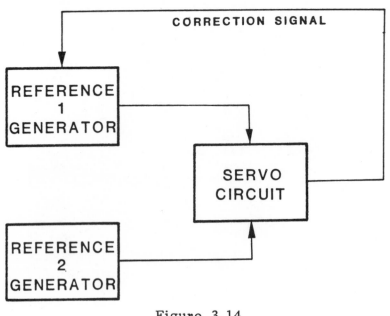

Figure 3.14
Servo Circuits

In a quality video tape recorder, there are usually separate servo circuits used to control the speed of rotation of the video head drum and the capstan. One reference signal used by these servo circuits is switched between the input video signals (during recording) and played back signals. The other reference signal comes from motion detectors attached to the head drum or the capstan cylinder.

Figure 3.15 shows that the drum servo generates a correction signal by comparing pulses filtered out of video input signal and pulses generated by the drum's motion detector. The motion detector senses position and rotational velocity of the drum. If the signal sensed by the motion detector indicates a slow-down of drum rotational speed relative to the frequency of vertical scans of the stable video signal, the servo circuit's correction signal commands that more power be applied to the motor to speed it up. (The start of the vertical scans are denoted by pulses embedded in the video signal and can be filtered out. These "vertical sync pulses" are discussed in greater detail in the next chapter.)

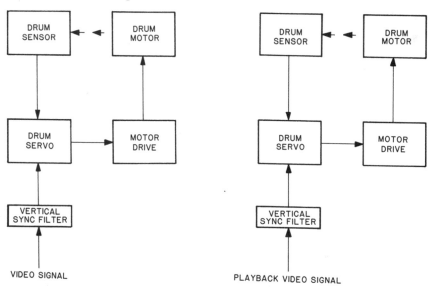

a) Record   b) Playback

Figure 3.15
Drum Servo Operation

During playback, the drum servo references the played-back video signal to command the drum to speed-up or slow-down to match the frequency of drum rotation (revolutions-per-second) to the frequency of video vertical scans (pulses-per-second).

The motion detector on the drum also determines when the switch between the two video heads is scheduled to occur. Most 3/4-inch recorders place this "head switch" during the vertical blanking interval. As shown in Figure 3.16, some older 3/4-inch recorders are designed so that head switch occurs between five and eight horizontal scans before the end of a vertical scan. This type of machine creates a picture in which the head switch shows up at the bottom of the picture as a few lines of horizontally displaced picture information. Usually this head switch is concealed by the mask around a picture tube's display area.

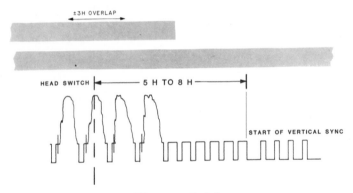

Figure 3.16
Head Switch Position

To maintain a constant speed reference for the tape as it moves across the transport, the filtered video signal pulses are recorded on a "control track" along the edge of the tape. These recorded pulses form a track of "electronic sprocket holes" that are used as the reference for control of the speed of the tape during playback.

The capstan servo references the filtered video pulses of the input signal during record to ensure that the pulse positioning during record is constant, as shown in Figure 3.17. During playback, the control track pulses played off the tape are compared to the input video signal or a stable pulse reference generated within the recorder. If it takes too long for the next control track pulse to arrive, the capstan motor speeds up to try to read the pulse as soon as possible. Should the control track become damaged or unusable, the capstan speeds up so that the picture appears to be in a slightly fast scan. (On some VTRs, loss of control track commands a switch so there is no playback picture and playback sound is slightly speeded up.) The constant comparison and correction between timing of the unstable played-back pulses and the stable reference pulses commands speed of rotation changes in the capstan (that in turn controls the speed of the tape).

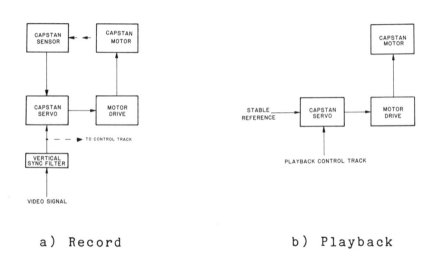

a) Record   b) Playback

Figure 3.17
Capstan Servo Operation

Notice that both servo circuits depend on a reliable and constant source of input video signal. Should the input video source be switched between two sources whose vertical scans do not start at *exactly* the same time, the time between the pulses denoting the start of vertical scans is different and causes the servo to detect an error and command correction.

For example, if a stable video signal is input to the VTR, the servo circuits become accustomed to operating with (approximately) 1/60-second interval between vertical sync pulses. If a non-synchronous video signal were switched to the VTR input, there can be between zero and 1/30-second between the vertical sync pulses surrounding the time of the switch. If the time between vertical sync pulses is too short, the servo circuit senses that the speed of the drum and capstan are too slow (because of the speeded-up reference) and commands that they speed up. Once the 1/60-second pattern of the second signal is established, things return to normal. The changes in servo response is seen in the picture as alternating expansions and contractions of picture size, similar to the picture breathing in and out. This is the way in which a picture monitor displays time base error.

TRACKING

The *path* of the spinning video head in the play mode must exactly overlay the *path* that the video head took in the record mode. As shown in Figure 3.18, if the path of the head in play does not exactly overlay the path of the head in record, the head detects interference between the magnetic patterns of two adjacent tracks. The interference between the two magnetic patterns of recorded information produces a display similar to that shown in Figure 3.19. To adjust this interference out of the picture, a TRACKING control is usually available on the video tape recorder to command the drum motor to temporarily slow down or speed up and adjust the video head path in play to overlay the path that the video head took in record. Once the motor places the heads in position, the servo circuits command the motor to maintain that positioning.

A TRACKING METER similar to that shown in Figure 3.20 is frequently designed into a video tape recorder to indicate the optimum positioning of the video head over the recorded tracks. This meter actually measures the amount of information being picked-up off the tape, with the maximum strength of signal indicated when the head path correctly straddles the track.

152

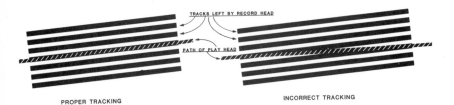

PROPER TRACKING   INCORRECT TRACKING

Figure 3.18
Tracking

Figure 3.19
Interference Bar in the Picture
from Tracking Misadjustment

Figure 3.20
Typical Tracking Meter

Some recorders have automatic tracking systems (not to be confused with the slow-motion "Automatic Tracking systems described later in this section). These systems record inaudible tones in the video signal. During playback, the tones are carefully monitored for any change in frequency or phase. Tone changes cause the circuits to command tracking changes for correction.

Video tape expands when it is exposed to extreme heat and contracts in extreme cold. Small changes in tape temperature (less than about 30°) between recording and playback are not noticeable, but large differences in temperature may record the path of the video heads of a properly adjusted transport in a tape location where a properly adjusted transport may be unable to read. The relocated video tracks may fall outside the tolerances specified for the tape format, making proper playback impossible with the TRACKING control adjusted to its limits. Maintenance of a constant tape temperature between record and playback (or editing) considerably reduces tracking and tension problems.

Some VTRs have special systems that allow playback at transport and head-to-tape speeds different than the speed in record. These Dynamic Tracking or "Automatic Tracking" systems allow creation of still-frame, slow-motion, and reverse-action special effects from the video tape. Dynamic Tracking systems consist of special video heads mounted on a "piezoelectric" material that deforms its shape on application of an electrical signal, a sensitive tracking-error detector circuit, and special switching and processing circuits. With this equipment the recorder is able to repeatedly track a single tape track for still frame, overlap tape tracks for slow motion and reverse action, and skip tracks for fast motion.

TENSION

Video heads are not in constant contact with the tape because the tape does not completely encircle the video drum. While the head is not in contact with the tape, the video drum experiences less drag and speeds up. As the protruding head first resumes contact with the tape and increases drag, it momentarily slows down the drum. This slow-down causes the head-to-tape speed to be slightly reduced until the drum servo circuits can catch up.

In most helical scan video recorders, the initial head-to-tape contact occurs at the start of a vertical picture scan. A tension error is usually seen as a "hooking" or "flagging" at the top of the playback picture because that is when the head's speed differential is greatest.

The momentary increase in head-to-tape drag happens both in record and play, but does not produce noticeable picture distortion unless the amount of slow-down is different between record and play. As shown in Figure 3.21, the amount of slow-down depends on how tightly the tape is wound around the video head drum. In the record mode, this tension is automatically maintained to an amount adjusted by a technician. The technician uses tension gauges similar to those shown in Figure 3.22 when making his adjustments.

In the play mode, a front-panel SKEW or TENSION control allows operator adjustment of tape tension. This control may be used to correct the playback signals for recorder tension misadjustments, variations in tape temperature and humidity, and tape drag variations that are sometimes found in cassette housings. The TENSION control is adjusted to eliminate any bend at the top of the picture.

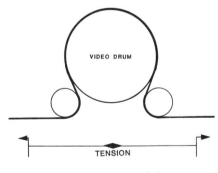

Figure 3.21
Tape Tension

Courtesy Tentel® Corporation
Campbell, California

Figure 3.22
Tension Meter Measuring Video Tape Tension

TAPE DAMAGE

The spinning video head can encounter other problems if a tape becomes creased or folded. A physical deformation of the tape moves the tape closer or farther away from the head. As shown in Figure 3.23, if the distance from the source of the magnetic field (the head in record and the tape in play) increases, the strength of the magnetic field decreases. Thus a crease may produce weaker magnetic coupling with the tape resulting in a weaker playback or recorded signal and possibly a complete loss of signal.

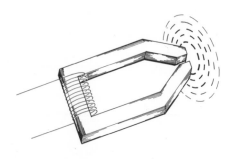

Figure 3.23
Distance from the Head
Affects the Strength of the Magnetic Field

A crease also increases the potential for physical damage to the fragile video heads as they speed through the damaged area. Many tape formats enclose the tape in a cassette housing to reduce damage, but this doesn't eliminate tape creases caused by transport malfunction or misadjustment.

In "small format" 3/4-inch, 1/2-inch VHS, 1/2-inch Betamax®, or 8mm VTRs, an arm in the VTR pulls the tape out of the cassette housing and wraps it past the erase head, audio heads, control head, and around the video head drum. Any time the recorder is in play, record, or pause mode, the tape is positioned across the transport and the cassette is locked into the recorder. For 3/4-inch recorders in the stop, fast forward, and rewind modes, the tape is kept within the cassette housing and shuttled directly between the supply and take-up reels.

The A, B, and C "medium formats" that use 1-inch wide tape and "large format" machines that use 2-inch tape were designed around open reel transports. Some large format machines also offer optional cassette packaging of tapes.

Courtesy The Dub House

Figure 3.24
Typical 1" and 2" VTRs

157

Additional tape and video head protection is provided by an "end-of-tape" circuit. As shown in Figure 3.25, a lamp and a light-sensitive diode are mounted on the transport so that the tape moves between them as it winds across the transport. As the tape passes between the lamp and photocell, a clear plastic leader at the beginning and end of a tape allows the light to pass through. When the oxide-coated portion of the tape (which is opaque) does not block the light between the lamp and the photodiode, the recorder does not operate in the play or record modes. End-of-tape circuits may also operate using light reflected off the oxide coating to sense tape presence.

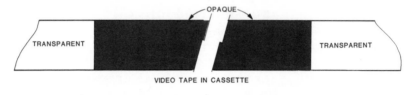

VIDEO TAPE IN CASSETTE

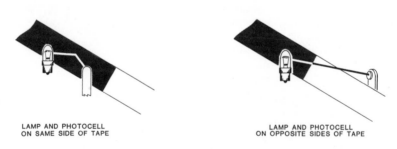

LAMP AND PHOTOCELL
ON SAME SIDE OF TAPE

LAMP AND PHOTOCELL
ON OPPOSITE SIDES OF TAPE

Figure 3.25
End-Of-Tape Detection

PAUSE MODE

Many video tape recorders have a "pause" mode to allow stopping the tape in place on the transport without the tape winding back into the cassette housing. For short periods of time, operation in the pause mode displays a still picture. This indicates that the tape is still in contact with the video heads and the video head drum. If a normal tape is left threaded around the spinning drum for an extended period of time (more than about 30 seconds), enough friction and heat can be generated by contact with the video drum to cause permanent damage to both the tape and the video heads.

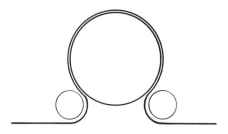

NO TENSION

Figure 3.26
Pause Mode

To alleviate the problems of pause, many VTRs offer a "long pause" or "extended pause" mode that leaves the tape threaded across the transport, but relaxes the tape tension around the video head drum. In this mode, there is little, if any, tape-to-head contact (and no playback picture is available). But it does allow the tape to be poised for immediate use over a long period of time.

The problems of pause can also be reduced by using video tape specifically intended for "still frame reproduction," as marked on the tape box. This type of tape has a special long-lasting formula of lubricant and oxide "glue."

COLOR VIDEO RECORDING TECHNIQUES

DIRECT RECORDING

Medium and large format VTRs that use 1" and 2" video tape, as shown in Figure 3.24 and 3.27, usually have the capability of "direct recording" an entire NTSC color video signal without a significant amount of signal processing. The signal recorded on the tape is a frequency modulated (FM) version of the entire NTSC video signal that is input to the recorder. When a "low" video signal voltage is present (representing "dark" picture details), a "low" frequency (7.9 MHz) FM signal is created. When a "high" video signal voltage is present (representing "bright" picture details), a "high" frequency (10.0 MHz) FM signal is created. A sync pulse in the input video signal creates a "very low" frequency (7.06 MHz) FM signal. During playback, this FM signal is demodulated (or "tuned") in much the same way that an FM radio tunes audio signals. Direct recording formats generally offer eight-to-ten generations before significant picture degradation occurs.

Courtesy Ampex Corporation

Figure 3.27
One-Inch Video Tape Recorders

HETERODYNE RECORDING

Small format recorders, those using 3/4-inch and narrower tape, must use other signal processing techniques in order to properly record and playback an acceptable color signal. Figure 3.28 shows several recorders that use this special "heterodyne" processing technique where two signals of different frequencies are combined to achieve color hue correction.

160

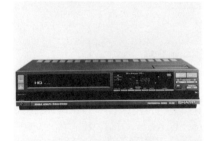

Figure 3.28
Heterodyne Video Tape Recorders

Even with the stabilizing actions of servo circuits, the video signal played back by a small format VTR is still not stable enough to allow accurate reproduction of a color picture. Multiple-generation recording is a special problem with the heterodyne recording formats generally offering three-to-four generations before significant picture degradation occurs. With the writing speed and tape standards of the 3/4-inch VTR, a 0.00000007 inch error in tape position produces a noticeable 5° subcarrier phase error.

To comply with the strict chrominance phasing requirements that are required for a high-quality picture, small-format VTRs first filter the color information (which has a frequency of 3.579545 MHz) from the luminance (black-and-white) component of the encoded video signal to allow special processing. The chrominance is processed to allow detection and correction of the chrominance phase errors that remain after undergoing the stabilizing effects of the servo circuits. Preparation for this color correction is automatically accomplished during the recording of the video signal, leaving phase correction to take place during playback.

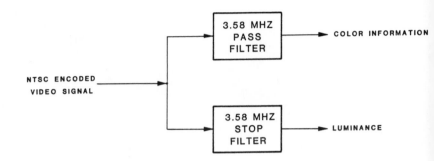

Figure 3.29
Heterodyne VTR Input
Color Correction Circuits

Reviewing the phase correction process in detail, the chrominance is first separated from the luminance of the video signal and then combined with a constant signal (4.267 MHz in small format VTRs) to produce a third 688 kHz signal that is the difference in frequency between the signals being combined.

$$4.267 \text{ MHz} - 3.579 \text{ MHz} = 0.688 \text{ MHz} = 688 \text{ kHz}$$

As shown in Figure 3.30, this "heterodyned" chrominance is then added to the frequency modulated luminance portion of the video signal before being recorded on the tape by the video head. Figure 3.31 shows the spectrum of signals applied to the video head.

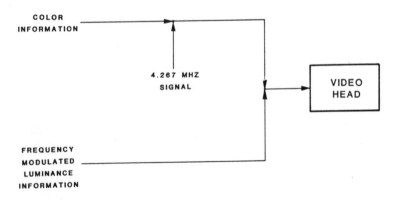

Figure 3.30
Heterodyning Chrominance Information

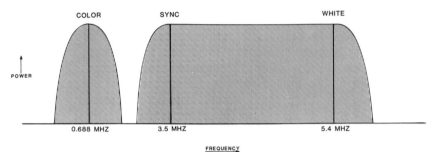

Figure 3.31
"RF" Signal as Applied
to the Video Recording Heads

Figure 3.32 shows that heterodyning two signals creates two output signals that are equal in frequency to the sum and the difference of the frequencies of the original signals. In 3/4-inch VTRs, the 7.846 MHz sum resultant signal is filtered out, leaving the 688 kHz signal to carry the color information. The results of heterodyning can be easily seen if the reader wishes to compute and combine the instantaneous values of two sine waves.

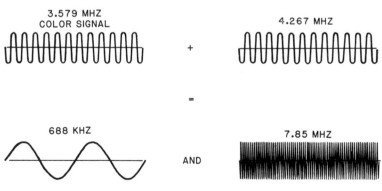

Figure 3.32
Heterodyning Chrominance

During the record process, the 3.58 MHz chrominance is separated from the luminance information and heterodyned to 688 kHz. In playback, chrominance phase correction takes place when the 688 kHz color information is again heterodyned with the VTR's 4.267 MHz signal to regenerate the 3.579 MHz standard.

$$4.267 \text{ MHz} - 0.688 \text{ MHz} = 3.579 \text{ MHz}$$

When the mechanical process of taping produces even a slight change in the frequency of the chrominance signal, a phase change results. The amount of frequency error of the chrominance signal from the tape can be expressed in the following formula:

$$4.267\text{MHz} - (0.688\text{MHz} + f_{error}) = (3.579\text{MHz} + f_{error})$$

As shown in Figure 3.33, the playback video signal's burst (with constant phase and amplitude) is used as the reference to determine the amount of color phase error. This 688kHz signal with phase error is heterodyned with a stable 3.579 MHz signal to create a signal of 4.267 MHz with phase error. A stable 4.267 MHz signal that is internally generated is phase compared with the playback 4.267 MHz signal with error. This comparison creates a signal of 4.267 MHz with error to be combined with the playback 688 kHz with error signal to create an "NTSC-type" video output signal.

With this phase correction process, a stable color signal is generated regardless of frequency, time, or phase error produced by the recording and playback processes. While the heterodyne frequencies that have been discussed are applicable to the conventional 3/4-inch and Betamax® formats, processing using other heterodyne frequencies are also used in the 3/4-inch SP, VHS and 8mm formats. Usually, use of higher frequencies require video tape of special formulation for efficient use. The heterodyne frequencies for many video tape formats are listed in Appendix B.

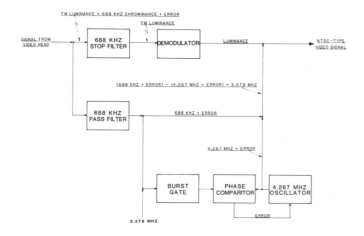

FM LUMINANCE + 688 KHZ CHROMINANCE + ERROR
FM LUMINANCE

SIGNAL FROM
VIDEO HEAD

| 688 KHZ STOP FILTER | → | DEMODULATOR | LUMINANCE | NTSC-TYPE VIDEO SIGNAL |

(688 KHZ + ERROR) − (4.267 MHZ + ERROR) = 3.579 MHZ

| 688 KHZ PASS FILTER | 688 KHZ + ERROR |

4.267 MHZ + ERROR

| BURST GATE | PHASE COMPARITOR | 4.267 MHZ OSCILLATOR |

3.579 MHZ

ERROR

Figure 3.33
Heterodyne VTR Output
Color Correction Circuits

Y/C RECORDING

If the luminance and chrominance (Y and C) portions of the video signal are kept separate during multi-generation recording operations, the picture can be greatly improved. Super-VHS and ED Beta (Extended Definition) 1/2-inch formats have been developed to capitalize on the picture improvement that can be realized by using separate circuitry and connectors for Y and C. Y/C equipment uses basically the same heterodyne format as regular VHS and Betamax equipment. Although two separate cables are necessary to properly interconnect the Y and C video components, Y/C equipment is not a true "component" format (discussed in the next section), but is an improvement over existing heterodyne formats.

As discussed in Chapter 1, the horizontal scanning rate was changed from 15,750 Hz to 15,734 Hz and the color signal was standardized at 3.579545 MHz to reduce interference between chrominance and luminance. Despite this effort, some interference remains. This remaining interference is the primary cause of generation loss in video tapes. If the luminance and chrominance components of the video signal are kept separated in the taping processes, the generation loss is minimized.

Courtesy JVC Professional Products Company

Figure 3.34
Typical Super VHS Equipment

Luminance/Chrominance interaction shows up in the picture as moving dots around high–contrast details. This "dot crawl" can be considerably reduced if luminance and chrominance are kept separate through the recording processes. The chrominance of Y/C formats is processed in exactly the same fashion as conventional heterodyne formats.

Y/C formats record the luminance and heterodyned chrominance on the video tracks on the tape in almost the same manner as any other heterodyne recording format. To achieve maximum performance, the Y and C signals must be processed in separate circuitry and separate cables must be installed. To interface with existing equipment, an NTSC-type encoded color signal is available to be recorded by conventional recorders or for viewing on a conventional picture monitor.

COMPONENT RECORDING

Recorders using true component recording techniques eliminate interaction between chrominance and luminance by separating the luminance and two chrominance components (R-Y and B-Y) in three different locations on the tape. Component recording formats generally offer five-to-six generations before significant picture degradation occurs. Sony's BetaCam® and Matsushita's M-II, as shown in Figure 3.35, are component recording systems.

Luminance is separated from chrominance, frequency modulated in much the same manner as direct and heterodyne systems, then recorded on one diagonal track of the tape. The chrominance information is separated into its R-Y and B-Y axes before being recorded on another diagonal track of the tape. (One video head is used for luminance and a second, adjacent video head is used for chrominance.) The R-Y and B-Y signals are two different combinations of the original red, green, and blue video signals that are within the chrominance portion of the encoded video signal.

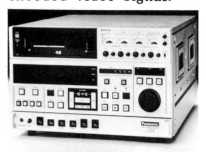
Courtesy Panasonic Broadcast Systems

Courtesy Panasonic Broadcast Systems

Courtesy of Sony Corporation

Courtesy of Sony Corporation

Figure 3.35
Typical Component Recording Equipment

As shown in Figure 3.36, BetaCam® systems time-compress the two chrominance components to occupy the same video head scanning time as spent by the uncompressed luminance signal. During playback, the compressed chrominance components are expanded back to their original times. M-II equipment operates in much the same manner.

□ = Y

▨ = B-Y

▦ = R-Y

HORIZONTAL SCAN NUMBERS ARE SHOWN

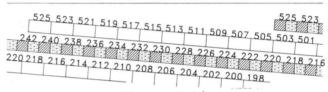

Figure 3.36
BetaCam® Recording Format

DIGITAL VIDEO RECORDING

The recording of digital codes that represent the video signal represents a considerable improvement in multi-generation recording over the analog recording systems that were discussed in the preceeding sections. (A discussion of digital concepts is presented in the next chapter.) The Digital Video Recorder (DVR) or Digital Television Tape Recorder (DTTR) shown in Figure 3.37 uses 19mm-wide tape to record up to approximately sixty-four generations of video tape before any significant picture degradation can be seen. This number of generations can be multiplied if lesser quality pictures are acceptable.

The television picture of Figure 3.38 illustrates the production technique allowed by a DVR. Each of the waves of water are successive generations, the motion of which were matched with digital video effects to each other. The bridge itself is a one-generation "layer" that is action-matched to the motor scooter. This layering production technique can save a considerable percentage of the production budget if correctly utilized.

Courtesy of Sony Corporation

Figure 3.37
Typical Digital Television Tape Recorder (DTTR)

Courtesy Limelite Video

Figure 3.38
Photograph of a Digital Television Picture

The technique of digital recording is technologically very difficult to accomplish. Sophisticated processing is used to detect any digital code abnormality and, once detected, correct or conceal the distortion that would appear in the picture. Using these sophisticated techniques, even dramatic recording problems as severe as a single head clog are corrected and virtually invisible to a viewer. Details of this "D-1" format are presented in Appendix B.

AUDIO RECORDING TECHNIQUES

The audio sections of many video recorders operate in exactly the same fashion as conventional audio tape recorders. As shown in Figure 3.39, the audio signal is added to a non-varying "bias" signal (to reduce the number of octaves that must be recorded) and sent to the recording head. During playback, the bias portion of the signal from the head is discarded, leaving only a representation of the original audio signal.

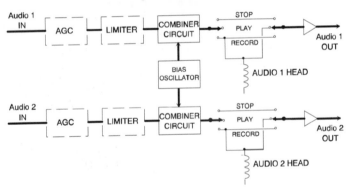

Figure 3.39
Audio Recording Circuits

In "Hi-Fi" models of VHS-format recorders, the audio is first frequency modulated, then recorded by drum-mounted audio heads at a layer that is very deep in the tape oxide. After the audio is recorded, the video heads on the same drum record the processed video signal at a layer that is very shallow in the tape oxide. The track of the heads across the tape is similar, but the differing depths of recorded information help separate the video from audio (like different altitudes of aircraft on the same airway). In

170

addition to the layer separation of the signals, the audio heads are aligned with a different "azimuth" (*magnetic* direction of recording) than the video heads.

In Hi-Fi Betamax® recorders, the audio is frequency modulated (AFM) with a carrier frequency that falls between the heterodyned chrominance and the lower frequency of the modulated luminance. This combined AFM and processed video is recorded and played back by the video heads.

Some recorders offer very high quality audio recording and playback using digital signal processing techniques. (Digital signals are discussed in detail in the next chapter.) In those recorders, the benefits afforded the video signal in a DTR are available for the audio as well. Frequently the term "Pulse Code Modulation" (PCM) is used to describe this digital audio technique. Using the PCM technique, the digitized audio signals are frequently placed in otherwise unused times of the video signal (like vertical blanking).

Two audio circuits frequently found in video tape recorders, automatic gain circuit (AGC) and limiter, are discussed in detail in Chapter 2.

TAPE COPYING

To make a single copy of a previously recorded tape requires two machines, one to play back the tape and one to record a new copy of the tape. These two video tape machines are usually interconnected as shown in Figure 3.40. Figure 3.41 shows that multiple copies of a tape may be made at one time by connecting one playback machine and as many recording machines as required tape copies. The distribution amplifiers take their single input signal and outputs several independent but equal output signals. There is a more comprehensive discussion about distribution amplifiers in the next chapter. The copying process consists of playing back the signals in their original baseband video and audio form and rerecords them onto another tape. Notice that this rerecording process creates another generation loss of taped material.

171

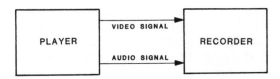

Figure 3.40
Block Diagram of a System
for Making One Tape Copy

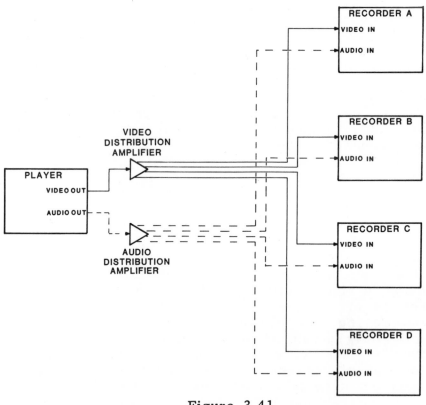

Figure 3.41
Block Diagram of a System
for Simultaneously Making Several Tape Copies

DUB MODE

Another method of making higher-quality video tape copies uses the "dub" mode of compatible heterodyne VTRs. The dub mode allows copying of taped picture information while bypassing many of the processing circuits normally used to play back the baseband video and audio signals. Considerable improvement in the picture quality produced by the recorded tape can be achieved if the noise and distortion effects of this signal processing can be bypassed.

As shown in Figure 3.42, the signal out of a VTR's DUB OUT connector bypasses the heterodyne circuits that convert the 688 kHz VTR color signal (in 3/4-inch VTRs) into standard baseband video 3.58 MHz subcarrier have been bypassed. Separate pins of the DUB OUT connector offer demodulated luminance and chrominance at 688 kHz (instead of 3.58 MHz). In the record VTR, the DUB IN connector bypasses the circuits that normally heterodyne the 3.58 MHz standard chrominance signal into a 688 kHz signal. This technique is called the "Y/688" mode. (Y=luminance and 688=chrominance in the signal. Notice the similarity in concept to Y/C recording.)

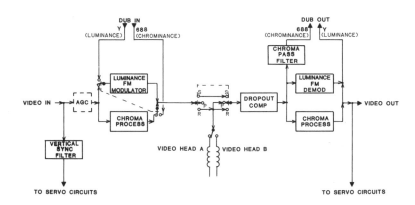

Figure 3.42
Dub Mode

An obsolete "dub mode" standard does not demodulate the luminance signal. "FM Dub" essentially, uses the combined signal directly read off tape (FM luminance and heterodyned chrominance).

Other VTRs offer "Color Dub" operating modes that bypass the playback phase correction circuits. Newer units offer a "Color Process" switch to bypass the circuits. Although the playback picture has unstable colors, a rerecorded signal has less distortion because of the missing processing. When the rerecorded tape is played back, the phase correction circuits correct for the total phase error introduced by the original play machine (with "Color Process" OFF to bypass the circuits) and the rerecord machine (with "Color Process" ON).

A VTR's "dub mode" should not be confused with "audio dub", which allows recording of new audio on one audio channel of a tape without altering or recording over any existing video or remaining channel audio. In the audio dub mode, the video erase head and the record heads are turned off while the selected audio channel is in the record mode. In this manner, tapes can be dubbed after production so the two audio channels can carry two languages, special coded signals for machine control, or two separate audio messages.

POST-PRODUCTION

If the scenes in the original tape(s) after production need to be rearranged, post-production editing equipment is available. During post-production, the original electrical signals may be altered to look and sound better or meet specifications. These "post" processes are handled electronically, and (unlike most audio and film editing processes) do not result in pieces of tape "on the cutting room floor."

Using an "editing bench" wired as shown in Figure 3.43, the original taped production material is placed on a tape player (or a recorder in the "play" mode) and rerecorded on another recorder. As shown in

Figure 3.44, at the end of the first desired scene, the recorder is stopped and the player is cued to the beginning of the next desired scene. Once the player is cued, the recorder and player then start and the next scene is rerecorded. This process continues until all the scenes on the "edit master" tape in the recorder are arranged as desired.

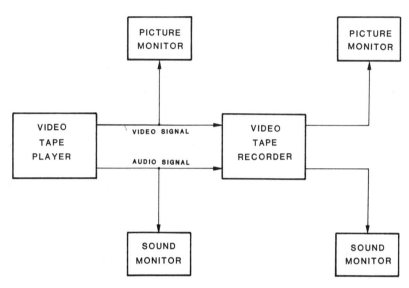

Figure 3.43
Block Diagram of a Basic Editing Bench

Actually, additional equipment is needed to rapidly complete the rearrangement job. First, a special "edit controller" or "editor" commands the player and recorder to automatically perform the edits at the proper times. The edit points of the tape are defined by the operator. An "edit-in point" defines the starting time of an edit. An "edit- out" point defines the ending time of an edit. As shown in Figure 3.45, an editor accepts up to three edit points among player edit-in, player edit-out, recorder edit-in and recorder edit-out points. The fourth point is automatically computed by the editor to prevent an attempt to squeeze in too much time from the player into too short a time slot in the recorder or trying to put too little time from the player into the designated time slot in the recorder.

SCENES AS THEY WERE RECORDED
ON THE PRODUCTION TAPE

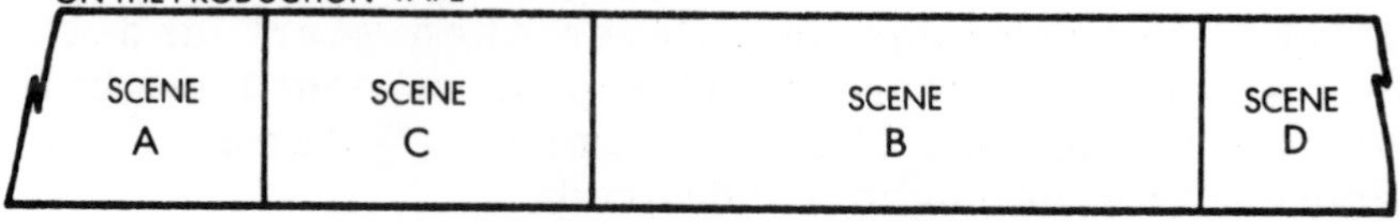

SCENE A	SCENE C	SCENE B	SCENE D

EDITING STEP	PLAYER MODE	RECORDER MODE
1	PLAY SCENE A	RECORD SCENE A
2	FORWARD TO SCENE B	STOP
3	PLAY SCENE B	RECORD SCENE B
4	REWIND TO SCENE C	STOP
5	PLAY SCENE C	RECORD SCENE C
6	FORWARD TO SCENE D	STOP
7	PLAY SCENE D	RECORD SCENE D

SCENES, AS RECORDED ON
THE EDIT MASTER

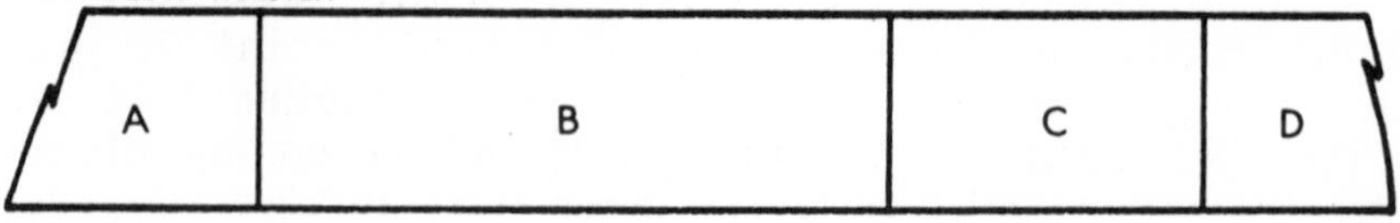

A	B	C	D

Figure 3.44
Editing

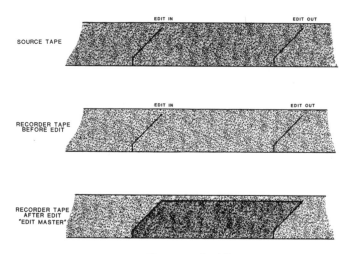

Figure 3.45
Edit-In and Edit-Out Points

With the capability of editing video and (normally) two audio channels simultaneously, monitoring equipment is necessary to preview and select scenes and sounds. Monitoring is connected across the recorder inputs (or source outputs) to see exactly what scene and sound is available for rerecording. Additional monitoring is connected to the recorder outputs to see the finished edit in context with the preceeding and following scenes and sounds.

The idea of cameras switched into a single recorder is the classical idea of multiple camera video production. With increased portability, increased reliability, and decreased equipment prices, many directors have decided on methods of multiple camera production that delegate an "iso" or "isolation" recorder to each camera. The system shown in Figure 3.46 "locks" the recorders together using a special "time code" signal. The program is then assembled entirely in post-production by choosing the desired viewpoint of the scenes during editing. This procedure ensures a high-quality final program when covering unrehearsed presentations and is frequently able to reduce labor-intensive production time. When editing this type of material, or when using multiple program sources or tapes for incorporation into a single edited master tape, the terms "A-Roll", "B-Roll", "C-Roll", etc. are used to refer to the source VTRs.

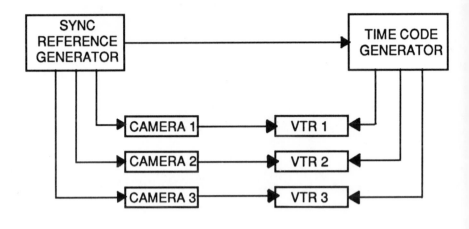

Figure 3.46
Block Diagram for a Multiple-Roll System

Additional equipment may be added to an editing bench to obtain a higher quality finished product. A processing amplifier ("proc amp") or time base corrector ("TBC") is a common addition to an editing bench. This equipment electronically improves the video signal from a tape to produce a stable and consistent quality picture. (A detailed discussion about TBCs and proc amps occurs in the next chapter.)

Audio can be "sweetened" to enhance the sound from production tapes. Equalizers, mixers, and other audio signal processing equipment may be used to enhance the aural scene in much the same way that TBCs and proc amps may be used to enhance the visual scene.

The basic editing bench described so far constitutes a minimal configuration for consistent quality finished products. Other options like character generators ("TV typewriters"), video switchers, audio tape players, special "key" cameras, or even more video tape players are commonplace in complex facilities. Figure 3.47 is a block diagram of a more complex editing facility.

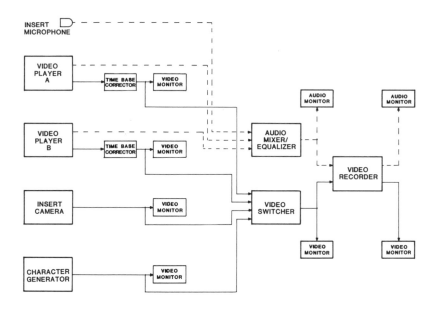

Figure 3.47
Block Diagram of a Complex Editing System

Not all video tape recorders can fulfill the special video and audio signal editing requirements of the post-production process. Special switching of video, audio, and control head functions; a "flying" video erase head (mounted on the drum assembly beside the video record/play head); a separate audio erase head; and extra-sensitive servo circuits must be provided to have a fully operational editing recorder.

Courtesy JVC Professional Products Company   Courtesy Panasonic Broadcast Systems

Figure 3.48
Typical Editing Recorders

To perform edits on a recorder, it is best if the recorder makes the transition during the vertical blanking interval (the time used for the bottom-to-top retrace of the picture scan). As shown in Figure 3.49, performing the edit when no picture information is visible does not allow a field composed of half old scene/half new scene to flash on the screen. Although this abnormal picture lasts no longer than 1/60-second (one field), it still presents an objectionable flash in the picture.

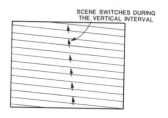

 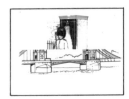 

Figure 3.49
Vertical Interval Editing

To accommodate this smooth picture transition, the vertical blanking interval of the video signal that is being played back by the recorder must coincide with the vertical blanking interval of the new video signal that is to be recorded. The servo circuits of an editing recorder are designed to handle such a chore.

Even if the vertical interval is used for the transition point of the edit, there can still be a distortion in the picture if the incoming video signal fields do not coincide exactly with the video signal fields of the played back video signal. In many editing recorders, a special "frame servo" makes sure that the edit transition occurs *only* between an odd field and an even field or between an even field and an odd field (never between two odd or two even fields).

As shown in Figure 3.50, if the last video signal field on the tape before an edit is an even field, and the first video signal field after the edit is another even field, the point where 262½ horizontal scans have been completed (bottom-center of the scanned picture) of the scene before the edit matches with the second horizontal scan path (top-left of the scanned picture) of the scene after the edit. Notice the difference (equal to one-half a horizontal scan path) between the end of the first scene and the beginning of the second scene. Such a scanning error forces a picture monitor to display a "whip," "flash," or "flag" at the top of the picture during the first field of the new scene. Once the picture monitor has a chance to catch up with the new timing of the video signal, the picture returns to normal. Usually this takes less than one-half the field (about 1/120-second).

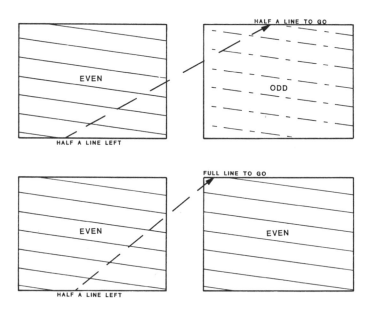

Figure 3.50
Flag Edits

Significant flagging problems in editing operations can occur any time that there is a drastic difference in the timing of the scan between the previously recorded scene and the new scene. Before an edit is performed to a signal from another VTR, the TENSION or SKEW controls on the source VTR should be adjusted for minimum flagging at the top of the picture. This ensures that there is a minimum amount of horizontal positioning error between the scan positions at the bottom of the previously recorded scene and the top of the new scene. If available, a cross-pulse monitor (discussed in Chapter 1) may be used for critical examination of the flagging picture.

Most editing recorders allow the operator to select between two editing modes:

ASSEMBLE EDIT - Entirely new program material.

INSERT EDIT - Drops new program material into previously recorded material.

As shown in Figure 3.51, after receiving a command from an editor to change from play to record, an editing recorder in the assemble edit mode changes the operating mode for all the signal tracks *and the control track* at the next vertical blanking interval in the video signal. Just before the edit, the servo circuits command motor speed changes to synchronize the pulses played back from the control track of the tape on the editing recorder with the vertical interval of the incoming video signal. In this way, the edited transition occurs during the vertical blanking interval of both signals. After the edit, the servo circuits lock to the incoming video signal and a new control track is recorded on the edit master.

In the insert edit mode, operator-selected tracks (Video, Audio 1, Audio 2, etc.) change from play to record, *but the control track remains in the play mode* (and the recorder servo circuits reference the previously recorded control track). Before an insert edit, the servo circuits synchronize the control track with the incoming video signal (in the same manner as the assemble mode). After an insert edit has been entered, the capstan servo circuits do not change reference and continue to synchronize the played back control track with the incoming video signal. Such control of coincidence of the vertical intervals of the recorded and new signals throughout the edited sequence allows smooth scene transitions both on entering and exiting the edited scene. Note that if there is not an existing control track, the recorder does not have a stable tape speed reference and unstable servo operation results.

VTR Editing modes can now be redefined to mean:

ASSEMBLE EDIT - *New Control Track* is recorded.

INSERT EDIT   - *Existing Control Track* is used and retained as new Video is being recorded.

If one were to compare the visible results of the two types of edit modes, one sees:

ASSEMBLE EDIT   - The recorder is stable on entry, creating a clean entry point, but as the recorder creates a new control track, it loses coincidence with the old control track. The exit point produces distortion. On exiting the scene, a band of noise passes through the picture because the full-tape-width effect of the erase head is experienced. (Because a given point on the tape passes the erase head before the video and audio heads, a gap of unrecorded tape is created if all recording operations are simultaneously stopped.)

INSERT EDIT   - The recorder uses existing control track and is not disturbed because speed and reference points are maintained. The entry and exit points are both stable and undistorted. Erase heads for individual audio and video channels are used to eliminate the noise band experienced during an assemble edit exit.

At the operator's option, the insert mode may be used in all editing activity (including "assembling" a program by simply adding on program segments) by first making a normal video recording to record a control track on the fresh tape, then performing the insert edits. Although any video signal can work, a solid-field black or gray picture signal does not show a patterned flash (like a flash of color bars) should the edit not be performed correctly. This "laying black" technique also repacks the tape for more consistent VTR tension control during editing.

Both of the editing modes require precise alignment between the incoming video signal and the video signal as read off tape. A "pre-roll" time is programmed into automated editing operations to allow the recorder servo circuits and mechanics of the VTR(s) reach normal operating speed. An editor commands the VTRs to back-up two to fifteen seconds playing time before an operator programmed edit point and stop (or "park") before beginning the actual edit procedure. The more unstable the playback signal, the longer the pre-roll time required for a clean edit.

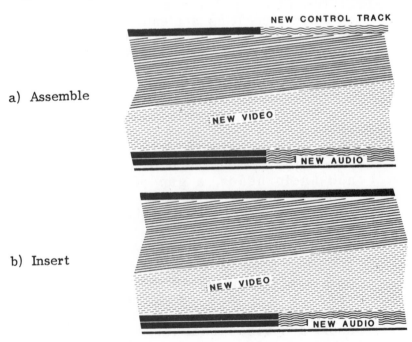

a) Assemble

b) Insert

Figure 3.51
Edit Modes

EDITORS

To ease the operator's burden of controlling two or more video tape units during editing, control units have been developed to reduce some of the more tedious tasks. A video tape "editor" provides remote control of all the normal player and recorder operating modes (play, record, fast forward, reverse, search, pause, etc.). Besides the manual control functions, automatic functions such as pre-roll, edit, and post-roll sequenced operations are designed into the editor to accomplish multi-task editing operations with ease. Some editing systems provide multiple edit information storage and are able to command the operation of video switchers, audio mixers, and audio equalizers for scene-to-scene optimization and matching.

Most editors have the capability of "previewing" an edit before it is actually recorded, then return the recorders to the appropriate pre-roll starting point. After previewing an edit, the operator has the option of recording the edit as entered, or changing ("trimming") the edit points. Each editor is different and has different displays, functions, and controls; but they all have a few functions in common.

Courtesy of Sony Corporation

Courtesy Panasonic Broadcast Systems

Figure 3.52
Typical Editors

The editor counts pulses played back from the control track or a special "time code track" or "address track" to determine scene locations and editing points on the tape. A counter is provided to display the position of the player and recorder tapes (in "minutes:seconds:frames" of time).

With a control track editor, the operator "zeros" the counter after the tape has been threaded. As the tape is shuttled during editing operations, the editor counts control track pulses to display positioning relative to the zero point, as shown in Figure 3.53. Remembering that the tape is removed from control track head contact and replaced into the cassette during stop, rewind, and fast forward, the tape could be moved without any pulses being detected to describe a change in tape position. Any tape shuttling or a change in tape loading cannot be detected or displayed by a control track editor and an error in the exact edit point may occur.

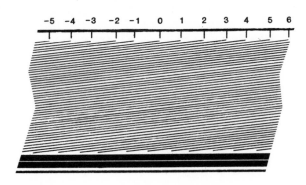

Figure 3.53
Setting Zero of a Control Track Editor

TIME CODE

Some sophisticated editing operations use a code representing absolute tape positioning that is recorded on the tape before beginning editing operations. The recorded "time code" signal is composed of digital words that identify each frame in much the same way as edge numbers identify each individual frame of motion picture film.

Time codes are usually recorded on a special address track on the tape (as shown in Figure 3.54 for

3/4-inch VTRs) or during the vertical blanking interval of the recorded video signal. When an address track or an audio track is used, the technique is called "longitudinal time code" (LTC). When the code is placed in the video signal, the technique is called "vertical interval time code" (VITC). As detailed in Appendix C, time code standards have been developed by the Society of Motion Picture and Television Engineers (SMPTE) and the European Broadcasting Union (EBU) to be recorded as a semipermanent record of tape positioning. The 80-bit LTC words are not directly compatible with the 90-bit VITC words although a computer may be used as an interpreter.

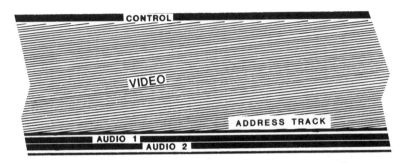

Figure 3.54
3/4" Video Tape Address Track

Instead of using the control track and losing the tape positioning reference every time a tape is changed (or the recorder or player is placed in stop, fast forward, or reverse), time codes allow positive reference of positioning among various points on several tapes simultaneously. The code is read off the tape, decoded, and the position displayed when the tape is moved to read a valid code sequence.

Time codes are particularly useful when there are many source tapes or for "off-line" editing. One-inch and digital editing benches are expensive to purchase and maintain, a factor reflected in their high rental rate. To reduce the cost of editing, off-line edit benches that use small VTR formats at a low rental rate are used to create a "work print" of a program. The work print is the result of most of the time-consuming editing decisions that go into the finished program.

As shown in Figure 3.55, the first step in using an off-line edit system is to pull a small-format copy of the original tapes that have a window of character generated time code information "burned in" the picture. The burn-in contains time code information in the hours:minutes:seconds:frames format. These tapes are then edited on the inexpensive off-line editing system. As the work print tape is edited, the burn-in shows the unique time codes at the beginning and end of each scene. These codes are logged into an "edit decision list" (EDL). The edit decision list is used to rapidly and easily conform the large format "edit master" tape.

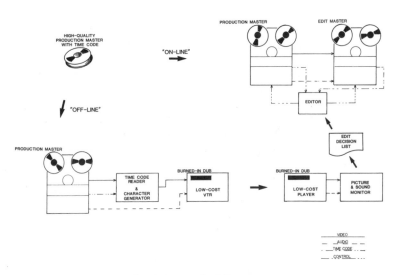

Figure 3.55
Off-Line Editing Procedures

Some off-line systems generate the edit decision list on a computer disk. When the computer disk is loaded in the editor controlling the on-line large-format system, the editor can automatically conform the edit master tape. Sophisticated editing systems may use the EDL to automatically control the start and stop times of picture dissolves, wipes, and special effects as well as special audio mixes and equalization.

188

Most simple editors perform command functions, leaving video and audio signals to travel directly between the two machines as shown in Figure 3.56; however, some editors provide video and audio signal control to fade or dissolve between scenes. These commands may be provided by a switch closure on a "General Purpose Interface" (GPI) or by intelligent commands via an "RS-232C" or "RS-422" connector.

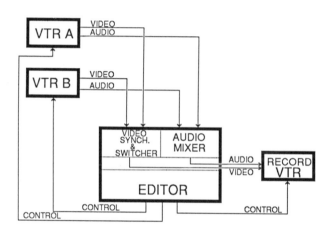

Figure 3.56
Control of Picture and Sound
by Sophisticated Editors

There are many manufacturers of VTRs and there are many other manufacturers of editors. It is not unusual to require a modification of a particular VTR to operate with a particular editor, although this practice is becoming increasingly rare. Such modification requirements should be carefully examined for reliability, manufacturer support in case of trouble, and invalidation of a manufacturer's warranty or support.

VIDEO TAPE

Although magnetic tape is usually some form of iron oxide (basically the same as ordinary rust) coated on plastic, the manufacture of magnetic tape is a complex art that produces wide variances in the quality of the finished product. One batch of tape may have outstanding qualities, while the next batch from the same manufacturer cannot stand up to the wear and tear of routine video operations.

The quality of the raw tape product from the various manufacturers is a frequent topic among video users. The manufacturer's overall quality control of the tape becomes the primary variable in tape quality.

The most important characteristics of video tape are oxide formulation, lubrication, adhesion of the oxide to the plastic base, tape life, and susceptibility to damage in use. Each of these characteristics has a direct relationship with final picture and sound quality as well as the usable life of the product.

DROP OUTS

To keep the oxide coating on the plastic base, a "glue" is responsible for the maintenance of a thin, even coating of oxide material. This binding must maintain the thin coating throughout the wear and tear of wrapping and unwrapping around a recorder's tape guides, drums, capstan, and other transport parts. If the glue fails, the oxide may flake off, producing a momentary loss of signal, or "drop out." A drop out is seen in the picture as a small area of brightness flash. The brightness of this small picture flash area may be determined by the adjustment of a "drop out level control inside the VTR."

Many VTRs, including most of the cassette-based units, have internal circuits or provision for external circuits that allow for "drop out compensation" (DOC) to mask the effects of most drop outs. If provision has been made for external compensation of drop outs, the VTR is equipped with an "RF OUT" or "DOC OUT" connector on the rear panel. As shown in Figure 3.57, the signal from this connector is an amplified version of the combined modulated luminance and heterodyned chrominance signal directly off the head.

190

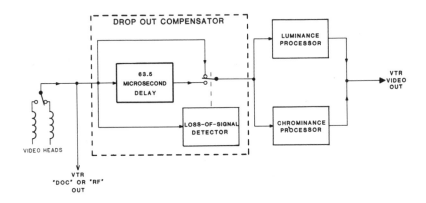

Figure 3.57
Drop-Out Compensation (DOC) Circuit

In most drop out compensation circuits, a portion of the off-tape signal is amplified and delayed by 63.5 microseconds (the time spent on one horizontal scan). The remainder of the off-tape signal is amplified and sent through a transistor switch to the luminance and chrominance processing circuits. If a loss of the signal is detected, as is the case during a drop out, the delayed signal is switched on the line for the duration of the dropout. A dropout rarely has a duration greater than the time spent during one horizontal scan.

HEAD CLOGS

If the flaked-off oxide that created a drop out happens to lodge in the gap between the poles of the video head as shown in Figure 3.58, a magnetic short circuit prevents any magnetic field from being generated or detected. This "head clog" does not allow any signal to be recorded on tape or played back from tape. Prevention of head clogs is the primary reason for maintaining a clean transport and operating environment.

191

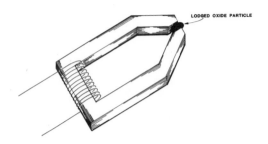

Figure 3.58
Head Clog

VTR CLEANING

Head clogging problems can be reduced by regular cleaning and proper maintenance of the recorder so that any particles that do flake off are not left in the recorder to eventually lodge in the head gaps. The cleaning of a video head should be accomplished with lint-free application of head cleaning solution. (To reduce the possibility of shattering the head by shock-cooling with head cleaner, the head should be allowed to cool if it has been recently in contact with the tape. If using a spray-type head cleaner, spray the cleaner on the swab or chamois instead of directly on the head.) The applicator should not leave any fibrous pieces, as is left by cotton or cloth. Foam-tipped swabs and chamois are popular cleaning materials. Many cleaning solutions for magnetic heads should not be used on rubber or plastic materials. The instructions found on the container of the cleaning solution should be strictly followed. In a pinch, alcohol (preferably denatured alcohol) may be used, but the water content in alcohol is so high that it tends to dry out rubber parts.

Several manufacturers provide "cleaning cassettes" that use abrasion to dislodge any particles in the gaps of the heads. Since this technique erodes the heads like sandpaper, such cleaning cassettes should be used sparingly - if at all - to maximize the life of expensive recording heads. The best cleaning cassettes use a liquid cleaner applied to a non-fibrous material loaded in the cassette housing and are advertised as being "non-abrasive."

TAPE LUBRICANT

Working in opposition to the glue between the oxide and base, a lubricant is mixed with the magnetic particles to reduce wear, and allow the tape to travel smoothly over the various surfaces of the recorder. An increase in lubricant can reduce the effectiveness of the glue to the point that the glue can lose its grip and allow the oxide to drop off as a fine powder. If not enough lubricant is present, excessive wear reduces the life of both the tape and the recorder.

DEW SENSOR

Even with the benefits of a lubricant, any moisture present on the spinning video head drum causes adhesion between the tape and drum. With the tape stuck, the drum motor may stall and burn out, the drum motor driving circuitry may burn out, the video head may break, and the tape may be destroyed. Most recorders now have "dew" sensors that do not allow operation when moisture is present. Recorders can develop condensation on the video head drum after being moved from a cool, dry environment (like air conditioned buildings) to warm, moist environments (like outside in the summer). If a recorder move is anticipated under those conditions, enough time should be allowed before operation (about an hour) for all the condensation to evaporate from the video head drum.

PHYSICAL TAPE DAMAGE

Another consideration in tape evaluation is the susceptibility of the tape to damage (like creasing or folding). This is a function of the lubricant as well as the base material. The thicker the base, the less susceptible the tape is to damage, setting a practical limit to the amount of tape that can be packaged in a given size cassette housing. For long life, the base must withstand the strain of ordinary use.

Aside from the tape itself, the housing that protects the tape should be evaluated for amount of drag generated by the tape reels and for wearability in all areas of the housing. A cassette housing should be examined for susceptibility of guide tab problems and for mechanical stability of the tape reels.

VIDEO DISCS

Technology is rapidly expanding in areas of video signal storage. No longer is magnetic tape the only option for storage and distribution of video material. Video discs similar to those shown in Figure 3.59 are emerging as a practical reality for mass distribution.
Optical recording is the only currently popular method of storing and retrieving signals on discs although capacitive methods, like capacitive electronic device (CED) and video high density (VHD) techniques, have been explored in the marketplace. As shown in Figure 3.60 when playing back a disc using the optical system, a laser shines a narrow light beam on the surface of the rotating disc. This surface has been etched to provide variations in light reflectivity to correspond to variations in the original video and audio signals. The laser light reflected off the disc varies in proportion to the disc's reflectivity to be detected by a light sensitive diode (photodiode) that reproduces the video and audio signals.

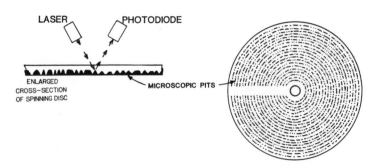

Figure 3.59
Optical Video Disc

The optical video disc itself has a series of microscopic "pits" that have been burned into the surface and then covered with clear plastic. These pits are burned into tracks around the disc so that one or two fields (depending on the operating mode) are read in each revolution of the disc. These tracks can be either in a spiral or concentric pattern, depending on the standard used. In normal operation, the disc rotates at 1800 rpm to generate all the information necessary for faithful reproduction of the video and audio signals.

In most disc applications, the original video production is taped and edited into final form. A copy of the edit master is forwarded to a disc manufacturer for "replication" (making an exact replica) on video disc.

Some video disc equipment has three laser systems to allow recording and an almost immediate playback. These "direct read after write" (DRAW) and "write once read many" (WORM) systems are usually more expensive than conventional tape technology.

The technology involved in recording and manufacturing an optical video disc is complex and expensive. Videodisc use is currently limited to program distribution where many copies are required. In this way, the high cost of set-up and generation of discs can be distributed among the many copies and users.

STILL-FRAME VIDEO

There have been several methods developed to record still video images. A popular low-cost method records still video images on small computer-like floppy discs. Called "still-frame video" or "electronic still photography," a standard recording format has been established to record up to 25 complete frames of video on a single disc.

Each disc comes in a 2" X 2" X 1/8" holder. On a given disc, there are fifty concentric tracks, each of which can contain one field ($\frac{1}{2}$ frame) of video information or up to twenty seconds of audio information.

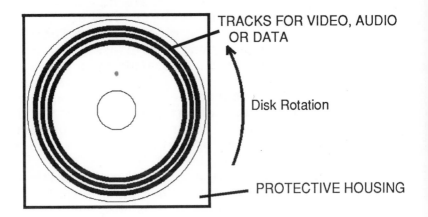

TRACKS FOR VIDEO, AUDIO OR DATA

Disk Rotation

PROTECTIVE HOUSING

Figure 3.60
Still-Frame Video Disc Format

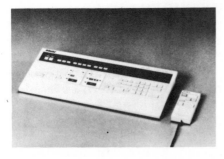

Courtesy of Sony Corporation

Figure 3.61
Typical Still-Frame Video Equipment

QUESTIONS

1. If the control track head does not read a control track during playback, does the capstan servo slow down or speed up?

2. If the picture breaks-up, what clue would determine whether the capstan servo or drum servo is at fault?

3. What field production procedures can be used to minimize tracking and tension problems?

4. Using a cross-pulse monitor, what would indicate proper adjustment of tension of a small-format helical-scan VCR?

5. On a 1/2" VHS-format tape (shown in Figure 3.13), what would the VTR do if there was severe damage to the upper edge?

6. Can longitudinal time code (LTC) be converted to vertical interval time code (VITC)?

CHAPTER 4

VIDEO
SIGNAL
CONCEPTS

The television signals distributed among the many independent pieces of equipment must adhere to a rigid set of specifications to maintain picture quality. The output signal from a camera must meet the input signal requirements of a recorder. The output signal from a recorder must meet the input signal requirements of a picture monitor. Unlike the matching of audio circuits described in Chapter 2, video signals have been standardized.

Because of the complexity of the video signal and the many different situations in which television production must be accomplished, everything must be exactly right to achieve the highest quality of signal. Sync generators, time base correctors, and processing amplifiers are used to support production and post-production activities and to correct minor distortions that the video signal encounters along the way. Waveform monitors and vectorscopes allow consistent monitoring of the signal on its own terms - voltage and time - instead of our human terms of picture brightness, position, saturation and hue.

SYNCHRONIZING SIGNALS

Whenever two pieces of video equipment need to interact, they must scan the image in step with each other. A video monitor must scan in step with the video signal from a camera or recorder, otherwise the displayed picture is jumbled (as though the VERTICAL or HORIZONTAL HOLD controls were misadjusted). As shown in Figure 4.1, a pulse is put at the exact time that each horizontal scan begins and at the time that each vertical scan begins. These synchronizing (or "sync") pulses are used to ensure that a focused image detail detected in the camera is in exactly the same relative place within the picture displayed on a monitor.

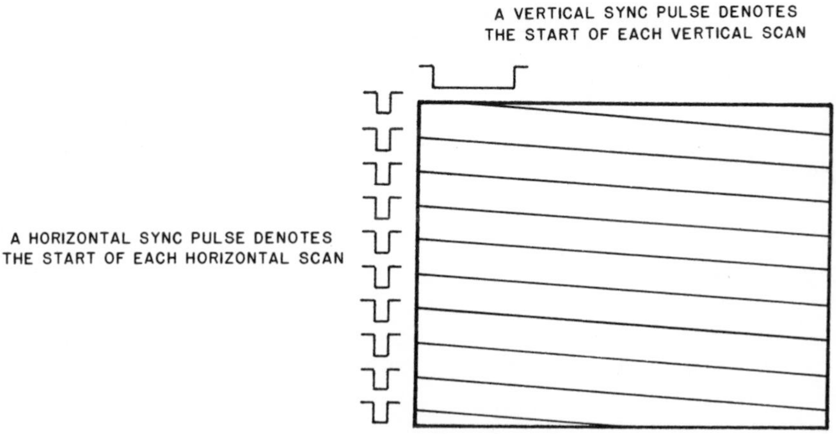

A VERTICAL SYNC PULSE DENOTES
THE START OF EACH VERTICAL SCAN

A HORIZONTAL SYNC PULSE DENOTES
THE START OF EACH HORIZONTAL SCAN

Figure 4.1
A Sync Pulse Commands
the Start of Each Scan

SYNC GENERATORS

A circuit inside the camera, or inside a separate piece of equipment, generates these sync pulses at the proper times. As shown in Figure 4.2, sync generator circuits usually have separate outputs for vertical sync pulses only ("vertical drive"), horizontal sync pulses only ("horizontal drive"), and a properly timed combination of horizontal and vertical sync pulses ("composite sync"). Properly timed "composite blanking" pulses are also generated to denote the starting and stopping times of picture brightness information during a scan. These pulses from a sync generator have a nominal 4 Volt peak-to-peak amplitude.

Color sync generators create additional signals to synchronize the generation of chrominance. These signals are discussed later in this chapter.

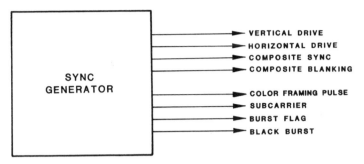

Figure 4.2
Sync Generator Outputs

SYNC PULSES AS SCANNING REFERENCES

Once a horizontal sync pulse is detected by either a camera or monitor, a horizontal scan of the picture is begun. After receipt of a horizontal sync pulse, the horizontal scan is continued until the next horizontal sync pulse is received. The scan is automatically completed by circuitry built into the monitor or camera.

The left-to-right scanning continues down the picture until a vertical sync pulse arrives. A vertical sync pulse marks the time to start each vertical scan. A circuit automatically completes the vertical scan after receipt of a vertical sync pulse. The horizontal and vertical deflection circuits operate independently of each other.

Once sync pulses are combined with the video (picture) and blanking information (VB) to form a "composite video signal" (VBS), both vertical and horizontal sync pulses have the same 286 mV peak-to-peak voltage amplitude. Circuits can be designed to separate horizontal and vertical sync pulses embedded together in a VBS signal. These circuits looking at two basic timing differences between horizontal and vertical sync pulses.

One basic difference between the horizontal and vertical sync pulses is the rate at which they occur. There are 59.94 vertical sync pulses every second and there are 15,734.266 horizontal sync pulses every second (in North American color systems).

The second basic difference is the duration of the pulses themselves. Vertical sync pulses consume the same amount of time as three entire horizontal scans.

VIDEO SIGNAL GRAPHS

Once the visual brightness information is converted to the video signal, the sync signals are often combined with the picture information into a composite VBS (Video, Blanking, and Sync) video signal that has *all* the information needed to make a picture. As shown in Figure 4.3, the sync and picture components of a composite video signal are timed so that all the information necessary for picture generation can be sent down one wire or through one transmitter.

The "blanking level" is the voltage to which the video signal returns during the retrace of the electron beam in a pickup tube or picture tube. The blanking level is what differentiates between sync (with signal voltages below blanking level) and picture luminance components (with signal voltages above blanking level.)

The "horizontal blanking interval" is the time allocated within the video signal to complete the retrace of the electron beam from the right side of the picture back to the left side. There are some special names applied to various portions of the horizontal blanking interval. First, a "front porch" is the time spent at the blanking level before a horizontal sync pulse (to the left of the pulse). After the horizontal sync pulse, a "back porch" of blanking

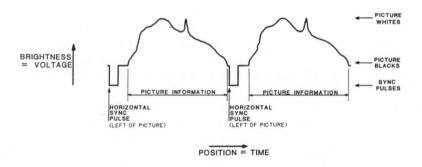

Figure 4.3
Graph of Two Horizontal Scans
of a VBS Signal

level occurs before the picture begins. Following the
composite video signal voltage graph shown in Figure
4.3 through the time spent to describe one field of
picture, the vertical sync pulse first arrives (in the
same video signal) to command the vertical deflection
circuits to start providing a varying voltage to the
vertical deflection coils or plates around the pick-up
tube (in a camera) or picture tube (in a monitor) to
"slowly" deflect the internal electron beam to scan the
image or picture area from top-to-bottom. Horizontal
sync pulses then arrive to command the horizontal
deflection circuits to start providing a varying voltage
to the horizontal deflection coils and deflect the
electron beam to "rapidly" scan from left-to-right.
After 262 horizontal sync pulses (to create half of the
525 horizontal scans in North America), another
vertical sync pulse arrives.

The "vertical blanking interval" is the time
allocated within the video signal to accomplish the
retrace from the bottom of the picture to the top.

SYNCHRONIZED SYSTEMS

In a system that uses more than one camera or
video source, each of the sources must scan in step
with each other, or be "synchronous." As shown in
Figures 4.4 and 4.5, if we were to switch between two
non-synchronous video sources, monitors and recorders
would sense a drastic instantaneous timing change in
the sync pulses. If such an abrupt timing change is
input to a video monitor, the picture rolls or tears
momentarily until the monitor becomes accustomed to
the new sync reference. An abrupt timing change
causes the drum and capstan servo circuits in a video
recorder to "hunt" until the servos lock-up the
rotation of the drum and capstan to the new timing
reference signal. Such hunting of the recorder servo
circuits causes temporary expansion and contraction
("breathing") of the picture until lock-up to the new
timing reference can be established.

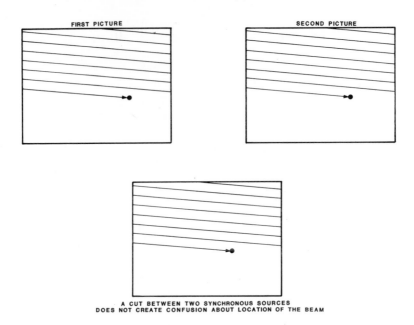

Figure 4.4
Switching Between Two Synchronous Sources

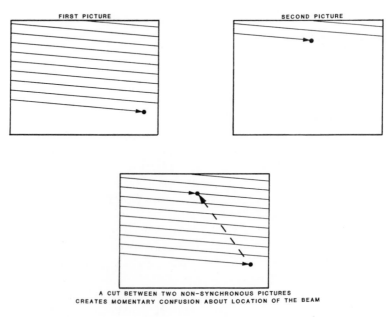

Figure 4.5
Switching Between Two Non-Synchronous Sources

There are three basic techniques that may be used to make equipment synchronous:

Line Lock - The 60 Hz AC power line is used as a stable vertical reference, leaving horizontal to either free run or indirectly lock to the power line frequency. (By government regulation, the power company must meet frequency standards that are acceptable to many television operations.) Because of signal standards explained later, line lock operation is available only in monochrome operation.

Sync Lock - Incoming horizontal and vertical sync pulses are used as references. Combinations of horizontal drive, vertical drive, composite sync, and composite blanking are typical signals required by sync lock equipment for proper operation. Color systems add subcarrier, burst flag or burst gate, and color framing pulses (discussed later in this chapter) to synchronize the color portion of the signals. Each type of synchronizing signal is sent through a separate coaxial cable.

Genlock - A composite (VBS) video signal with video and sync pulses is used by the equipment as the sync reference. The picture information is discarded and the "stripped" horizontal and vertical sync pulses are used as the scan timing references. The color bursts in the genlock input signal are used to synchronize the color signals, as discussed later in this chapter. A single coaxial cable can carry all the required synchronizing information.

VIDEO SIGNAL ANALYSIS

The video signal can be easily analyzed by examining a graph of detected brightness information. With white being positive and black being negative, examination of the graph shown in Figure 4.3 from left-to-right shows a horizontal sync pulse extending downward to super-black and surrounded by the black horizontal blanking level. Once the start of the scan has been established by the sync pulse, the graph shows the left-to-right picture information of the scene focused on the pick-up device. After the entire horizontal scan has been completed, another sync pulse arrives to denote the start of the next horizontal scan. This process continues until a large vertical sync pulse appears. If the voltage graph is started at a vertical sync pulse, after 262] horizontal scan paths (one field), another vertical sync pulse appears. The super-black vertical sync pulse is surrounded by the black vertical blanking interval.

Since the brightness of details in a scene determines the video signal voltage, the vertical axis of Figure 4.3 can be relabeled as "voltage." The amount of time spent in a horizontal scan (between horizontal sync pulses) is fixed, so the horizontal axis can be relabeled "time."

The location of a detail in the picture is defined by the amount of time delay between a sync pulse and the detail. Looking at Figure 4.6, assume that a picture detail is located exactly in the center of the picture. The voltage corresponding to that detail appears halfway (in time) between two successive vertical sync pulses to vertically define the position. The voltage corresponding to the detail appears halfway between the two successive horizontal sync pulses that define the start and stop of the horizontal scan that appears in the middle of the picture.

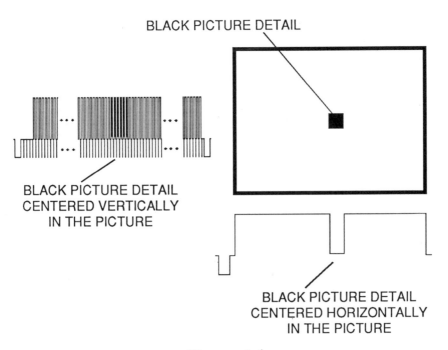

BLACK PICTURE DETAIL

BLACK PICTURE DETAIL
CENTERED VERTICALLY
IN THE PICTURE

BLACK PICTURE DETAIL
CENTERED HORIZONTALLY
IN THE PICTURE

Figure 4.6
Graph of a Picture Detail
Appearing in the Center of the Picture

WAVEFORM MONITOR

The voltage graph of a video signal can be displayed in real time by a special oscilloscope test instrument called a waveform monitor. Figure 4.7 shows some typical waveform monitors. The waveform monitor is indispensable to properly set up a camera, processing amplifier, time base corrector, and most other video signal handling and generating equipment. Video signal measurements are performed on the premise that the visual information is an electrical signal, not a viewable picture.

Courtesy Leader Instruments Corporation

Courtesy Tektronix, Inc.

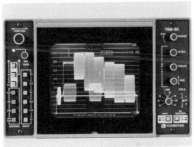

Courtesy Videotek, Inc.

Courtesy Tektronix, Inc.

Figure 4.7
Typical Waveform Monitors

INTERPRETING THE WAVEFORM

On a waveform monitor, the VERTICAL POSITION control is adjusted so that the blanking level (that occurs between the picture information of two successive horizontal or vertical scans) overlays the "0" (hash-marked) horizontal graticule line. As shown in Figure 4.8, with the blanking level adjusted to overlay the "0" line on the waveform monitor graticule, sync tips should fall to the "-40" line. All luminance information must then fall between "7.5" and "100" if signal quality throughout the television system is to be assured. (If a color signal is being displayed, most waveform monitors offer a LUMINANCE or IRE filter position to allow measurement of closely approximated luminance components of the video signal.)

210

Figure 4.8
Voltage Graph of One Horizontal Scan

The peak-to-peak amplitude of a composite video signal from sync tips at -40 units to picture whites at +100 units is equal to 1 Volt in baseband video (the "raw" video signal without having been modulated to a TV channel). The graticule units on a waveform monitor, established by the Institute of Radio Engineers, are called "IRE." In baseband television systems,

140 IRE = 1.000 Volt

100 IRE = 0.714 Volt

40 IRE = 0.286 Volt

1 IRE = 0.007 Volt

The limits of luminance voltage are inflexible. Any white picture detail above 100 IRE is frequently "clipped" off and ignored, to prevent overloading and distortion in monitors, recorders, and modulators. The picture produced by such a clipped luminance signal lacks details in the white or bright picture areas. On sources with many bright and sharp picture details (like white letters from a character generator), it is good operating practice for the sharp details to not exceed 80-90 IRE to reduce the sound "buzz" induced by some transmitters, modulators, and receivers.

Notice that with the blanking level adjusted to overlay "0" on the graticule of the waveform monitor, all luminance information is above "0" and all sync information is below "0." If the luminance voltage drops below the blanking voltage, it could be misinterpreted by a monitor as a sync pulse, creating confusion about the starting times of scans and causing the picture to jitter. The servo circuits in a video tape recorder receiving low luminance information become confused and "hunt" (speed-up and slow-down) while referencing between the real sync reference and the low video level. A 7.5 IRE "pedestal" of minimum black picture voltage is specified in the video signal to allow room for error as the signal passes through the system.

If the luminance signal from a picture is between 7.5 and 100 IRE and the picture looks good on a properly adjusted video monitor, the picture information is within acceptable limits. If the picture looks bad, the video monitor should be readjusted while displaying properly calibrated test signals.

VIDEO SIGNAL TIME BASE

As shown in Figure 4.9, when viewing a single horizontal scan of a black-and-white signal on a waveform monitor, the time (horizontal distance on the display) from sync pulse-to-sync pulse is the video signal "time base." In black-and-white systems, the time base is:

$$\frac{1}{\left(\dfrac{30 \text{ Frames}}{1 \text{ second}} \ X \ \dfrac{525 \text{ Lines}}{1 \text{ Frame}} \right)} = 0.0000635 \text{ seconds}$$

$$= 63.5 \text{ microseconds } (\mu s)$$

In color television systems, the time base is changed to eliminate interference with the color subcarrier frequency to 63.6 microseconds (μs).

212

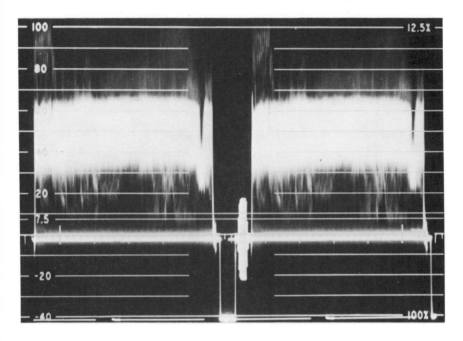

Figure 4.9
Waveform Monitor "2H" Display

Any variation from the standard in the time spent in a given scan is called "time base error." Any large variation in the time base shows in the picture as a roll, tear, movement, or momentary "glitch." Conversely, any time that a properly functioning picture monitor rolls, tears, or moves indicates a time base error. If the picture on the monitor recovers, the time base error was momentary.

It is also possible to adjust the horizontal axis of the waveform monitor to display video signals at the "vertical rate" (V or 2V) as shown in Figure 4.10, where the picture information from entire vertical scans is displayed. The horizontal sync pulses and picture information are squeezed together horizontally on the displayed waveform, leaving a pronounced vertical blanking interval and vertical sync pulse in the middle of the graph.

Figure 4.10
Waveform Monitor "2V" Display

Many waveform monitors offer the option of closely examining the vertical sync pulse with a "2V MAG" display as shown in Figure 4.11. When displaying the vertical sync pulse in this mode, one can see that the vertical sync pulse is not continuous, but is serrated with smaller pulses that appear at a rate that is double the horizontal sync pulse frequency. These small pulses are generated by the sync generator to ensure continued lock-up of horizontal deflection circuits in a monitor through the half line changes (remember that it's two hundred sixty-two *and one-half* lines per field) of scanning that occur on a field-by-field basis. (The serrated vertical sync pulse appears only in composite sync, not in vertical drive signals.) Double-frequency pre-equalizing and post-equalizing pulses are generated to occur both before and after the vertical sync pulse for the same reason.

214

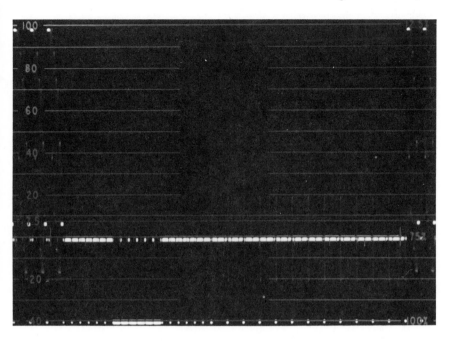

Figure 4.11
Waveform Monitor "Magnified 2V" Display

DC RESTORATION CONTROL

A switch on the waveform monitor that is labeled "DC RESTORATION" or "CLAMP" is used to stabilize the vertical position of the waveform under widely varying picture conditions. Normally, DC RESTORATION is left ON to smooth out any long-term voltage variations in the display.

If the DC RESTORATION is switched OFF, the waveform shifts up and down as picture content changes. While the DC restoration is turned OFF, field-rate video signal distortions like hum and bounce may be measured. Bounce shows up in a picture as unnatural brightness oscillations in the scene after an abrupt change in average picture level (APL). Hum shows up in the picture as horizontal bands of dark picture moving vertically through the picture. Since bounce and hum show up as variations in picture brightness, they are measurable on the voltage (vertical) axis of the waveform monitor.

215

WAVEFORM TRIGGER MODE

There is frequently another switch on the front panel that selects the sync source from which the waveform monitor "triggers" its sweep of the display. When switched to internal operation, the waveform monitor times the start of its sweep on sync pulses in the video signal being displayed. When the waveform monitor uses external sync signals, it displays the graph of one signal while starting its display sweep on command from sync pulses from another signal connected to the Sync Input on the rear panel. In the external sync mode, any time difference between the displayed signal and the external sync source is displayed as a horizontal shift in the graph. A displayed video signal that is not synchronous with the external sync source is displayed with either a horizontal shift or a horizontally unstable display. (Uses for the external sync display mode are discussed in the next chapter.)

In the 2H display mode, every other horizontal sync pulse is used to trigger a display sweep (because one 2H display takes the amount of time of two horizontal scans). The 2V display sweep is triggered to occupy the amount of time in two vertical scans.

Some sophisticated waveform monitors allow display of the video voltage from an individual scan. In these units, the trigger of the start of the display sweep is delayed until the same time (as derived from the sync pulses) from frame-to-frame.

The calibration of the horizontal axis of the video graph may be altered to look at individual portions of the signal. These special display modes may be used for measurement and quality assurance of sync pulses, VTR operation, and other distortions that may occur on a single scan of video information.

WAVEFORM CALIBRATION

Waveform monitors have an internally generated calibration signal. This signal is normally a square wave of 1 Volt peak-to-peak amplitude. When a waveform monitor is properly calibrated, the signal peaks at -40 IRE and 100 IRE. This signal is used for adjustment of the front-panel "GAIN" control on the waveform monitor.

The internally generated calibration signal should itself be calibrated to National Bureau of Standards time and voltage references by a certified repair station every eighteen months. (Government installations are required to undergo calibration every six months.) This calibration process assures accuracy and standardization of the display of signal waveforms.

SIGNAL STANDARDS

To ensure consistent operating results, strict technical requirements are imposed on the duration, shape, and levels of the various sync and blanking portions of a video signal. Current technical requirements of video signals of various organizations are detailed in Appendix E. The Federal Communications Commission (FCC) has standardized the requirements of video and audio signals *as broadcast*, but many individual television stations have adopted more stringent standards for incoming programming to ensure compliance to FCC standards after the signal has been processed and transmitted. These standards are also often adopted by non-broadcast facilities to maintain consistently high quality through their duplication and distribution efforts.

COLOR VIDEO SIGNALS

When the NTSC encoded color television signal was discussed in Chapter 1, hue and saturation of a color were said to be described by the timing and amplitude of a color subcarrier signal. Understanding the concept of phase (timing relative to a reference signal) is critical for understanding and measuring an encoded color signal.

SUBCARRIER PHASE

As shown in Figure 4.12, if two color subcarrier signals *with the same number of cycles-per-second (frequency)* are examined, the positive peaks and negative valleys of their voltage excursions may occur at exactly the same time. When this happens, there is no timing difference and the signals are said to be "in phase." But if one signal is delayed and the other signal is left alone, the peaks and valleys do not occur at the same time, changing the phase between the signals. The frequency of both signals has not changed from 3.58 MHz, just the relative timing of the voltage peaks and valleys between the two signals.

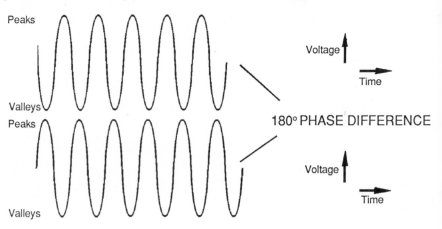

Figure 4.12
Phase Relationships Between Two Subcarrier Signals

The typical color video signal in Figure 4.13 shows how a reference "burst" of color subcarrier signal is placed in each horizontal blanking interval on the back porch (during the time between the horizontal sync pulse and the start of picture information. The time between the end of horizontal sync and the beginning of burst is sometimes called the "breezeway." As the picture completes a horizontal scan, the hues in the picture are described by comparing the phase between the "chroma" (subcarrier-frequency signal that describes the color of the picture details) and the burst reference.

The burst is constant in phase and peak-to-peak voltage amplitude. The picture information varies in phase (for picture hue) and in amplitude (for picture saturation), depending on the picture detail.

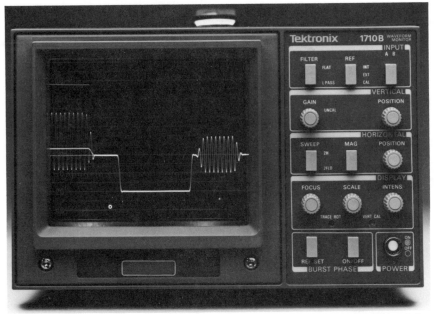

Courtesy Tektronix, Inc.

Figure 4.13
Waveform Monitor "1μs/div" Display

COLOR SYNC SIGNALS

A color sync generator provides the special signals shown in Figure 4.14 to ensure synchronous operation of the color circuits of the various pieces of equipment in a system. If the color is not synchronized, a color shift occurs when switching between two color video signals. In some advanced systems that require a strict relationship between color and sync, not only would the color shift, but there may be a shift in picture position or even a momentary glitch in the picture. (This is explored later in the Subcarrier-to-Horizontal Phase section of this chapter.)

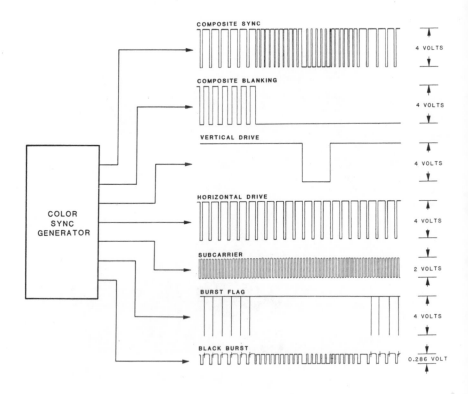

Figure 4.14
Color Sync Generator Outputs

The color sync generator outputs a continuous 3.58 MHz "subcarrier" signal with a 2 Volt peak-to-peak amplitude. This subcarrier signal is used by other equipment as the color standard for the system. "Burst flag" or "burst gate" pulses (with 4 Volt peak-to-peak amplitude) are generated to specify the time and duration after the horizontal sync pulse that the burst reference is to occur. It is easy to think of the burst as being created only when the burst gate is open to allow subcarrier to pass. (Actually the burst is 180° relative to subcarrier.)

Notice that the burst is not present on all horizontal scans during the vertical blanking interval. It is not present on scans numbered one through nine in each field. This corresponds with the duration of vertical sync, along with its pre-equalizing and post-equalizing pulses.

Many color sync generators also have a "black burst," "color black," or "crystal black" output that is simply the composite sync signal (at 0.286 Volts peak-to-peak) with the burst properly inserted. This black burst signal can serve as the synchronizing signal for equipment that can be genlocked. It is a signal that, when viewed, produces a black picture for it consists of a pedestal level (+7.5 IRE) with burst. Black burst signals are frequently used in genlock systems which synchronize equipment by using only one coaxial cable.

If the generator conforms to EIA RS-170A standards (discussed later in this chapter), a "color framing pulse" is output to positively identify one of the four color fields.

USING THE BURST

The detection of subcarrier phase relationships in the NTSC encoded video signal is complicated by the subcarrier reference burst signal not occurring during the entire scan. When the encoded color signal reaches a color picture monitor or receiver, the signal must be decoded into the original primary color signals. To properly decode the signal, the phase between the burst and chroma must be compared even though they occur at different times in the horizontal scan.

Each encoded color video monitor or receiver has a "subcarrier oscillator" circuit to generate a continuous subcarrier reference. As shown in Figure 4.15, the subcarrier oscillator signal locks to the phase of the burst reference signal to provide a regenerated subcarrier phase reference across the entire picture scan.

To adjust for hue errors in the picture, the phase between the burst in the video signal and the artificial reference generated by the subcarrier oscillator can be adjusted with the monitor's HUE or TINT control. This control changes the phase relationship between the unaltered chrominance information (that describes picture color detail) and the artificially generated subcarrier reference. As the reference signal changes, the *relative* phase of the picture details described by the chroma signal changes, and the hue of all the displayed color picture elements change by equal amounts.

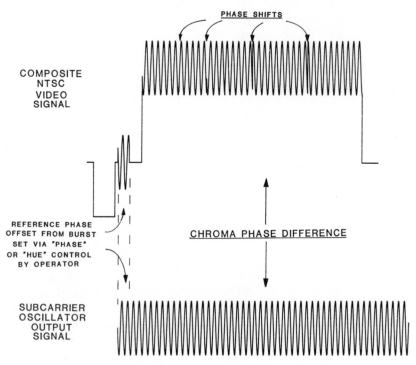

Figure 4.15
Subcarrier Oscillator Circuit in a Monitor
Phase-Locks to the Incoming Burst

VECTOR DISPLAYS

Accurate measurement of phase relationships between two signals can be difficult to accomplish. A waveform monitor can display the existence of phase relationships between subcarrier signals, but accurate measurement of such small time changes is impossible. To produce a measurable display of the color signals, the waveform monitor is replaced by a "vectorscope" or "vector display" similar to those shown in Figure 4.16. *Vectorscopes ignore everything in the video signal except the 3.58 MHz color subcarrier-frequency signals.*

Courtesy Tektronix, Inc.

Courtesy Tektronix, Inc.

Figure 4.16
Typical Vector Displays

To illustrate the need for a different measurement technique, one degree of color subcarrier phase difference is equal to 0.0000000008 seconds. Even with the one microsecond-per-division (0.000001 second-per-centimeter) scale that is the finest resolution normally available on a waveform monitor, one degree of color phase error is only 0.0008 centimeters (0.003 inch) wide. This small size, coupled with the inability to exactly define the voltage peaks and valleys, makes the measurement of color phase impossible on the waveform monitor.

VECTORS

A vector is a line with a specified direction and distance. As shown in Figure 4.17, distance from the center of the vector display corresponds to the peak-to-peak voltage amplitude of the color signal. This is equivalent to color saturation in the picture. This "radial" distance on the vector display is the same as the vertical peak-to-peak subcarrier amplitude displayed on a waveform monitor.

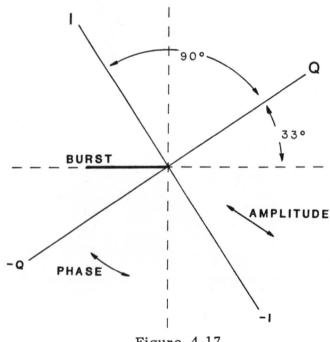

Figure 4.17
Vector Display Axes

VECTORSCOPE DISPLAY AXES

Phase is shown on the vector display as the angular distance around the circular "polar" axis. Remembering that the phase difference between the burst and the picture chrominance information is what determines picture hue, the burst is always placed at the nine o'clock position on the display. From this reference, all chrominance is measured as the angular displacement counterclockwise around the display.

Figure 4.18 shows a color bar signal displayed on a vectorscope. Notice that a bright dot in the display represents each color. The distance from the center of the display to any given dot represents the color saturation of that color bar. If we draw one straight line from the center of the display to the dot representing the burst, and a second straight line from the center of the display to our selected color bar dot, the angular distance between the lines (in the counterclockwise direction) represents the hue of that color bar. The two lines that we have drawn are really two vectors - each has a unique distance and direction.

For our purposes, the vectors in the display become measurements of signal amplitude (distance) and phase (direction) of the color information in a video signal. The vectors also represent picture saturation (chrominance amplitude) and picture hue (chrominance phase).

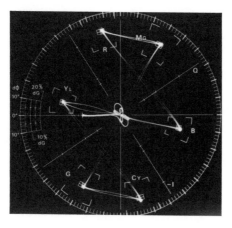

Figure 4.18
Vector Display of Color Bars

INTERPRETING VECTOR DISPLAYS

With the burst in the nine o'clock position, the I axis is 327° about the graticule and the Q axis is 327° - 90° = 237° about the graticule. These axes, as well as the R-Y (at the twelve o'clock position) and B-Y (at the three o'clock position) axes are etched on the graticule.

To further understand phase, take the red color bar as an example. From the encoder discussions in Chapter 1, we know that the NTSC encoded color signal uses the red (R), green (G), and blue (B) video signals to create luminance (Y), I, and Q signal components of red as follows:

$$\text{Red Y} = (0.30 \times R) + (0.59 \times G) + (0.11 \times B)$$

$$= (0.30 \times 0.714) + (0.59 \times 0) + (0.11 \times 0)$$

$$= 0.214 \text{ Volts}$$

$$\text{Red I} = (0.60 \times R) - (0.28 \times G) - (0.32 \times B)$$

$$= (0.60 \times 0.714) - (0.28 \times 0) - (0.32 \times 0)$$

$$= 0.428 \text{ Volts}$$

$$\text{Red Q} = (0.21 \times R) - (0.52 \times G) + (0.31 \times B)$$

$$= (0.21 \times 0.714) - (0.52 \times 0) + (0.31 \times 0)$$

$$= 0.150 \text{ Volts}$$

If the luminance signal component is ignored (as a vectorscope does), and the I and Q components are plotted on the vectorscope graticule in Figure 4.19, the dot representing red in the vectorscope display may be seen. To generate one definitive subcarrier amplitude and phase to represent the one red picture detail, a rectangle that has two sides composed of the plotted I and Q components can be used. One corner of

RESULTANT PHASE AND AMPLITUDE

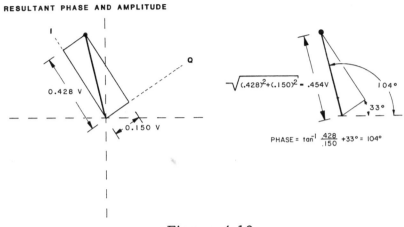

Figure 4.19
Vector Display of the Red Color Bar

the rectangle is adjusted to be the center point of the graticule, one corner at the maximum value of the I axis signal, one corner at the maximum value of the Q axis signal, and the fourth corner at the point representing the result of the vector addition of the I and Q signals (and the red picture detail).

Measurement from the graph reveals the red detail phase relationship to burst as 284° with a 0.454 Volt peak-to-peak amplitude. This result can be mathematically confirmed by using the Pythagorean theorem and the trigonometric relationships of right triangles.

Actually, the red color amplitude value changes to 0.451 Volts when the color encoding formulas are carried to several decimal points. Using the more precise formulas, the standard saturated color bars have subcarrier amplitudes and phase *relationships to burst* as shown in Figure 4.20. (To reduce interference during retrace of a horizontal scan, the burst is phased 180° to the applied subcarrier signal, changing the color bar subcarrier relationships to system subcarrier.)

227

COLOR	PHASE TO BURST (°)	CHROMA AMPLITUDE (Volts)
BURST	0	0.286
White	-	0
Yellow	347	0.319
Cyan	104	0.451
Green	61	0.423
Magenta	241	0.423
Red	284	0.451
Blue	167	0.319
Black	-	0

Figure 4.20
Color Bar Relationships to Burst

USING A VECTORSCOPE

The primary day-to-day operational use of a vectorscope is to measure subcarrier phase and amplitude adjustments made to color television equipment for consistent color picture rendition. Such consistency reduces picture color shift problems when changing from one camera to another, or when editing video taped scenes together that were recorded at different times or under different lighting or adjustment conditions.

Figure 4.21 shows that proper phasing of color bars using a vectorscope occurs when the burst is placed at nine o'clock using the vectorscope's PHASE control and the SUBCARRIER PHASE, CHROMA PHASE, or BURST PHASE control on the color bar source adjusted until the dots describing the color bars fall within the boxes etched on the graticule. The small boxes on the graticule provide a centered aim point for the large boxes (which represent the color bar tolerances allowed by the FCC).

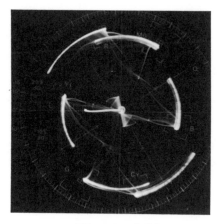

Figure 4.21
Phasing of Color Bars

Caution should be exercised to ensure that the vectorscope is adjusted so that the displayed dot is in the exact center of the graticule when there is no color in the input signal. Any time that the dot shifts from the center when the signal is input, there is subcarrier frequency information within the signal. When subcarrier is detected in a signal where no subcarrier should be present, the signal is said to contain "residual subcarrier" and the condition should be corrected by a technician to assure accurate color rendition.

BURST PHASE INDICATORS

A low-cost variation in phasing of color signals is provided by a "burst phase indicator" or "burst phase meter." In these units, proper phasing is indicated by colored indicators, as shown under the waveform display in Figure 4.22, or by indication on a meter.

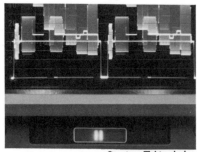

Courtesy Tektronix, Inc.
Figure 4.22
Typical Burst Phase Indicator

SUBCARRIER-TO-HORIZONTAL PHASE

Some sophisticated television systems demand a close correlation between the color subcarrier and sync. According to Federal Communications Commission (FCC) Part 73 technical specifications and recommended practice RS-170A of the Electronics Industries Association (EIA), the frequency of the color subcarrier is derived by the formula:

$$\frac{63}{88} \text{ X } 5,000,000 \text{ Hz } = 3,579,545.455 \text{ Hz}$$

The FCC and EIA also specify that the scanning of the television system be derived from the color subcarrier frequency as follows:

$$\text{Horizontal Scan Frequency } = \frac{2}{455} \text{ X Subcarrier}$$

$$= \frac{2}{455} \text{ X } 3,579,545.455 \text{ Hz}$$

$$= 15,734.26574 \text{ Hz}$$

$$\text{Vertical Scan Frequency } = \frac{2}{525} \text{ X } \begin{array}{c}\text{Horizontal Scan} \\ \text{Frequency}\end{array}$$

$$= 59.94005996 \text{ Hz}$$

Close examination of these specifications reveals a precise relationship between subcarrier and scanning. There are 3,579,545.455 subcarrier cycles-per-second. The scanning process completes one field in 1/59.94005996 second. These two signals interrelate to generate:

230

$$\dfrac{\left(\dfrac{3,579,545.455 \text{ subcarrier cycles}}{1 \text{ second}}\right)}{\left(\dfrac{59.94005996 \text{ fields}}{1 \text{ second}}\right)}$$

$$= 59,718.75 \dfrac{\text{subcarrier cycles}}{\text{field}}$$

COLOR FRAMING

The fractional number of cycles (xxx.**75**) in a field indicates that between the beginning and the end of a field there is a 1.0 - 0.75 = 0.25 cycle (90° phase) difference in the subcarrier. With this phase shift, returning to the original phase requires the completion of 1/0.25 = 4 fields. As shown in Figure 4.23, the phase shift produced when maintaining subcarrier-to-horizontal (SCH) phase effectively produces four fields in the NTSC system:

COLOR FRAME A:
> FIELD I starts after a line that is full of picture information and has a positive-going subcarrier at the leading edge of the sync pulse of Line 10 (where the burst first appears).

> FIELD II starts after a line that is half full of picture information and has a negative-going subcarrier (180° out of phase relative to FIELD I) at leading edge of the sync pulse of Line 10.

COLOR FRAME B:
> FIELD III starts after a line that is full of picture information and has a negative-going subcarrier at the leading edge of the sync pulse of Line 10.

> FIELD IV starts after a line that is half full of picture information and has positive-going subcarrier at the leading edge of the sync pulse of Line 10.

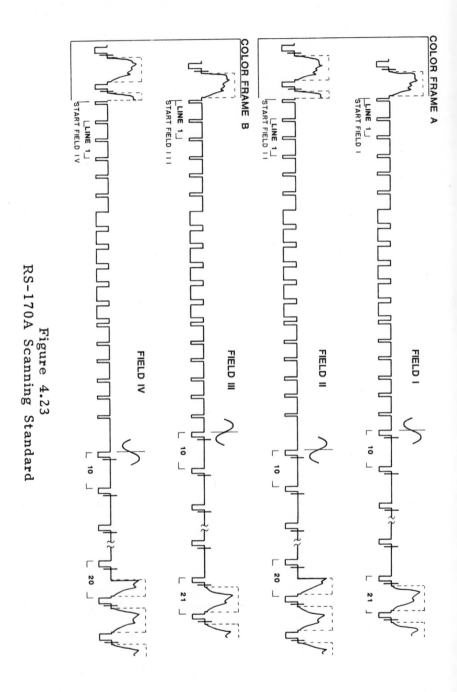

Figure 4.23
RS-170A Scanning Standard

FIELD	AT BEGINNING OF FIELD	
	TOTAL NUMBER OF S/C CYCLES	BURST PHASE
I	0.00	.00 X 360° = 0°
II	59,718.75	.75 X 360° = 270°
III	119,437.50	.50 X 360° = 180°
IV	179,156.25	.25 X 360° = 90°
I	238,875.00	.00 X 360° = 0°

Figure 4.24
Total Number of Subcarrier Cycles in a Color Frame

Examination of the totaled number of cycles at the beginning of each of the color fields confirms the four field interval that is necessary to return the subcarrier timing to the original phase. The number of subcarrier cycles that are counted starting at the beginning of Field I have fractional numbers to indicate a phase difference (relative to the start of Field I).

An edit between Field I and Field IV maintains the scanning standards, but stands a 50/50 chance of matching the color fields. As shown in Figure 4.25, if an edit were performed between Fields I and IV (with proper odd/even field scanning), the burst phase immediately after Field I should be 270° to properly match the burst phase at the end of Field I. But it instantaneously becomes 90°, which is 180° out of phase. This may produce a momentary color shift in the picture until the phase of the subcarrier oscillator in the video monitor has a chance to phase lock to the new reference. Of more importance, system equipment that is designed to maintain the SCH relationship may produce a momentary tear or roll in the picture as it instantaneously shifts operation by 140 nanoseconds (the effective time interval of a 180° subcarrier phase change).

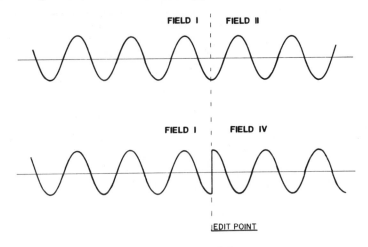

FIELD I | FIELD II

FIELD I | FIELD IV

EDIT POINT

Figure 4.25
Editing RS-170A Subcarrier Signals

SYSTEM SCH TIMING

Some professional video tape machines have circuits and servos that are capable of making edits while maintaining proper color field relationships. To derive maximum benefit from using machines that require proper SCH operation, three areas must be addressed within the television facility:

1) The sync must be generated with adherence to the standards established for SCH Phasing.

2) The entire television system must be timed to maintain SCH Phasing. (The details about system timing are discussed in the next chapter.)

3) The recorder and editor must be capable of recognizing and properly locking to an SCH signal.

The problems associated with maintenance of SCH through a television system can be substantial when one considers all the possible paths that can be taken by the video signal. It is not enough to simply maintain sync and video timing integrity; the subcarrier timing must also be precisely controlled.

COLOR FRAMING PULSE

Sync generators that are capable of generating sync and subcarrier signals with SCH phasing often provide a pulse for positive identification of the horizontal sync pulse of Line 11 of Field I. As shown in Figure 4.26, using this pulse for triggering an oscilloscope while displaying properly SCH-phased sync and subcarrier signals shows coincidence of the zero crossing of a negative-going subcarrier cycle and the 50% point of the leading edge of the horizontal sync pulse.

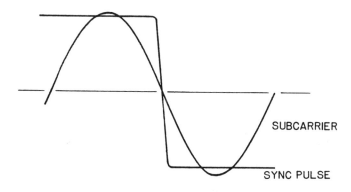

SUBCARRIER

SYNC PULSE

Figure 4.26
Subcarrier-to-Horizontal Sync
(SCH) Phase Relationship

DIGITAL AND ANALOG

There are two basic types of electrical signals used in the typical television facility - "analog" and "digital." Analog signals closely approximate the real world. Anytime that a real scene is viewed or a real sound is detected, an analog signal is generated; likewise, anytime that a real scene is to be displayed or a real sound recreated, an analog signal must be used.

Digital signals are artificial signals used in electronics systems to reduce the costs of signal storage or to reduce the effects of signal distortion.

Digital signals use precisely-timed pulses that vary between only two voltages to control equipment functions or to describe the picture information in a video signal. Analog signals use continuously variable voltages to provide the same functions. Each type of signal has its benefits and drawbacks.

One example of a simple digital signal is the output of a light switch. As shown in Figure 4.27, when the switch is on a relatively "high" voltage drives the light to full brightness. When the switch is off, there is zero voltage and the light is dark. There is no allowance for any brightness of the light (or the voltage driving the light) except full-on or full-off.

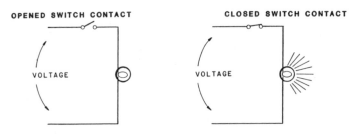

OPENED SWITCH CONTACT   CLOSED SWITCH CONTACT

VOLTAGE   VOLTAGE

Figure 4.27
Digital Output from a Light Switch

If we replace the light switch in Figure 4.27 with a light dimmer, as shown in Figure 4.28, an analog signal may be generated. The light is now capable of glowing at any brilliance between fully bright and fully dark, depending on the voltage output of the dimmer.

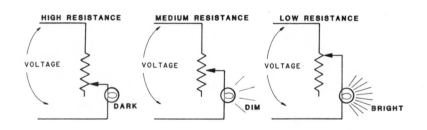

HIGH RESISTANCE   MEDIUM RESISTANCE   LOW RESISTANCE

VOLTAGE   VOLTAGE   VOLTAGE

DARK   DIM   BRIGHT

Figure 4.28
Analog Output from a Light Dimmer

If graphs are drawn as in Figure 4.29, with the vertical axis being the voltage being supplied to the light bulb and the horizontal axis being time, the difference between the two signals becomes obvious. There is no intermediate voltage on the digital signal, but the analog signal can have any intermediate voltage.

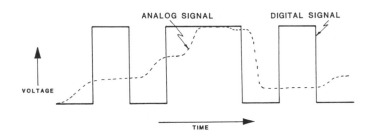

Figure 4.29
Voltage Graph of Digital and Analog Signals

USING DIGITAL SIGNALS

Every electrical system that carries or uses *either* digital or analog signals introduces distortions that alter the signal voltage amplitude. The content of the original digital signal can be "regenerated" from the distorted digital signal. As shown in Figure 4.30, the first step in digital signal regeneration is performed by circuits that detect the time that the distorted signal voltage passes through a "threshold voltage." These transition times are then sent to transistor switches to create a digital signal with the proper pulse amplitudes at the proper times.

A distorted analog signal cannot be conditioned to its original shape because the original signal amplitudes are usually not known and no single threshold voltage can be defined. The output of a given electrical system produces the same relative amount of distortion, whether the signal is digital or analog, but the ability to condition the digital signal back to its original shape and amplitudes suggests use of digital signals in critical applications.

237

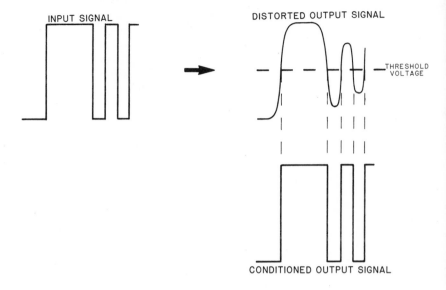

Figure 4.30
Regenerating a Distorted Digital Signal

Obviously, digital signals show great promise from an electrical point of view. But the microphone and TV camera sense information from an analog world. Not all sounds are the same loudness, nor are all scenes the same brightness. There must be some means of conversion from analog signals to digital signals, and conversion from digital signals to analog signals for effective use of digital technology.

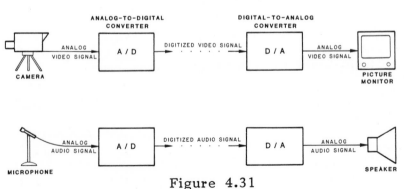

Figure 4.31
Analog Signals are Necessary
to Interface with the Real World

ANALOG/DIGITAL CONVERSION

Conversion from analog-to-digital signals (A/D) and digital-to-analog signals (D/A) can be easily accomplished. Using a noncomposite (without sync) black-to-white video "stairstep" signal in Figure 4.28 as an example, the A/D circuitry samples the voltages of the analog signal at precisely controlled times. The analog voltage sensed during those sampling times is sent to the digital coding circuits. In the example of Figure 4.32, the digital coding circuits have been designed with two "bits" (time slots) allotted to describe the information of each analog sample. Each bit is a precisely defined time slot in which the digital signal may be high or low. When a sample describes a low voltage (black) analog signal, the two bits available in the coding circuits are coded as No Voltage, No Voltage (Off, Off). Dark gray can be digitally coded as No Voltage, Voltage (Off, On). Light gray becomes On, Off. White becomes On, On. If this code is transmitted at the same rate and in the same sequence as the analog signal sampling, the analog signal is described by the digital code.

After a digital signal is received and conditioned to look like the original digital signal without distortion, a digital-to-analog conversion can be easily accomplished. Each unique code of two bits in Figure 4.33 command a bank of transistor switches to generate a unique analog voltage. When these analog voltages are generated in succession and at the proper rate, a closely *approximated* shape of the original analog signal results.

DIGITAL SAMPLING FREQUENCY

To describe the rapidly occurring voltage variations of an analog video signal, the analog voltages must be sampled at a very fast rate. Most systems use at least 11 million samples-per-second (equivalent to 3 or 4 times the color subcarrier frequency) for coding into digital form. This produces approximately 560 samples of the active picture in a given horizontal scan path and allows reliable sampling of the 3.58 MHz color subcarrier portion of the video signal.

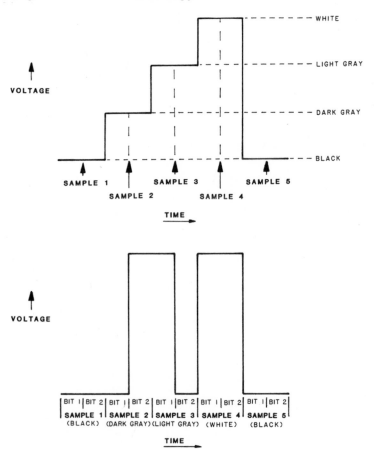

Figure 4.32
Two-Bit Analog-to-Digital Conversion System

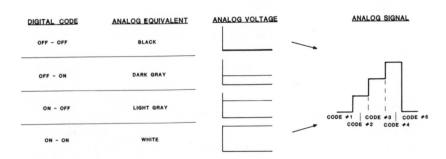

Figure 4.33
Two-Bit Digital-to-Analog Conversion System

240

In component digital video systems each of the video signal components is individually sampled. Most component digital equipment uses the ratio of 4:2:2 where luminance is sampled at a rate of four times the subcarrier frequency, R-Y is sampled at a rate of two times subcarrier, and B-Y is also sampled at a rate of two times subcarrier. The 4:2:2 relationship among sampling frequencies and subcarrier is one of several standards that have been internationally adopted in CCIR Recommended Practice Number 602. Composite digital systems are sometimes called "4:0:0" systems.

The sampling frequency of component digital video systems is referenced to subcarrier to assure internationally compatible equipment. The same basic equipment can be used for 525/60 NTSC systems, 625/50 PAL systems, and 625/50 SECAM. Only the input and output circuitry need be compatible with the particular analog standard. (Transcoding of the analog video signal standard is not facilitated by this standard; just the basic machine is made compatible among the various standards.)

DIGITAL CODE WORD SIZE

The infinite number of individual brightness levels of the analog video signal dictates that the coding be able to handle as many individual analog voltage levels as possible. The two bits of coded information in our example can only handle 2^2 = 4 brightness levels. Most composite digital video systems use eight bits of coded information to handle 2^8 = 256 individual brightness levels. Subjective viewing tests and video signal tests have indicated that a picture based on an 8-bit digital signal is of high quality.

DIGITAL SIGNAL APPLICATIONS

As discussed in the last chapter, digitized video signal recording provides a dramatic improvement over analog video signal recording systems. Because of signal conditioning and regeneration capability, the digitized video signal can endure tens of generations (copies of copies) of recording before significant visible picture degeneration occurs. This can be compared to analog technology where third or even second generation tapes encounter significant increases in noise and distortion. With this multiple-generation capability is exercised, many production options become economically feasible because many "layers" of a complex picture can be independently generated and manipulated.

In the typical television facility where analog video signals are primarily used, digitized video signals are commonly used in time base correctors and frame and field synchronizer equipment. As shown in Figure 4.35, these applications convert the incoming analog signal into a digitized video signal for easy storage in an array of individual semiconductor storage elements. Once a bit of the digital signal's information code is applied to a storage element, the element retains that information until another signal tells it to "write" in a bit from a new code. Since each bit requires a single storage element, an 8-bit code would require 8 storage elements to describe a converted analog video voltage. On receipt of a "read" command signal, the codes stored in the storage elements are released in the same order as they were written. The digital codes read out of the storage array are then sent to a digital-to-analog converter and an analog signal processing amplifier circuit.

Courtesy Fortel Incorporated

Courtesy Nova Systems, Inc.

Figure 4.34
Typical Time Base Correctors

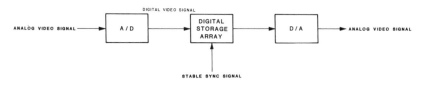

Figure 4.35
Block Diagram of a Typical Time Base Corrector

VIDEO SIGNAL PROCESSING

Processing amplifiers ("Proc Amps") and time base correctors ("TBCs") are used to correct some of the distortions that may occur during transmission or recording of an analog video signal. To correct for signal distortions routinely encountered, most pieces of video processing equipment (including TBCs) usually have individual controls for adjustment of the amplitude or phase of the burst, chroma, sync, and luminance components of the video signal. Some Proc Amps and all TBCs provide for complete regeneration of the sync and burst, reducing synchronizing problems associated when the signal has poorly shaped pulses and signals.

TIME BASE CORRECTORS

The ability to momentarily store a small amount of picture information in the TBC (as discussed in the previous section) is the only difference between a Proc Amp and a TBC. Most TBCs have all the controls and capabilities of a Proc Amp, plus the ability to correct for a time base error (where a horizontal scan does not take exactly 63.6 microseconds). The TBC momentarily stores the unstable incoming video signal and releases the information at a rate determined by pulses from a stable sync source. The TBC is frequently found correcting the output of a video tape recorder because of the extreme amount of time base error caused by the electrical and mechanical interplay in the recording and playback processes. A typical TBC interconnection schematic to a heterodyne VTR is shown in Figure 4.36.

243

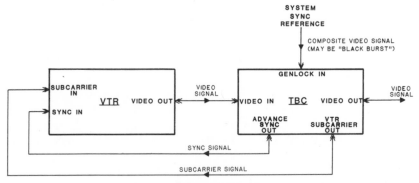

Figure 4.36
Interconnecting a TBC with a Heterodyne VTR

The total amount of time error that a particular TBC can handle is limited by the number of video signal storage locations or by the maximum amount of delay that can be introduced in the signal. This time "window" is frequently expressed in units of "H," the time spent in a horizontal scan. Figure 4.37 shows that the window is centered on the normal horizontal scan time, leaving half the window for any horizontal scans that take too little time and half for any horizontal scans that take too much time.

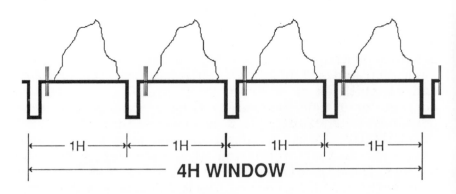

Figure 4.37
TBC Windows

PROCESSING AMPLIFIER CONTROLS

Processing amplifiers and processing circuits in a TBC usually provide the following controls for adjustment of analog video signal parameters:

PEDESTAL

The PEDESTAL or BLANKING control adjusts the minimum brightness voltage "base" on which the rest of the picture information is built (with the same visible results as BRIGHTNESS control adjustment on a video monitor). As is shown in Figure 4.38, the pedestal voltage determines the darkest black in the viewable picture and also serves as a safety factor in separating sync and picture information. As was shown in Figure 4.8, the PEDESTAL control of a Proc Amp should be adjusted to give a luminance pedestal voltage of 7.5 units above the blanking level ("0") as displayed on a waveform monitor.

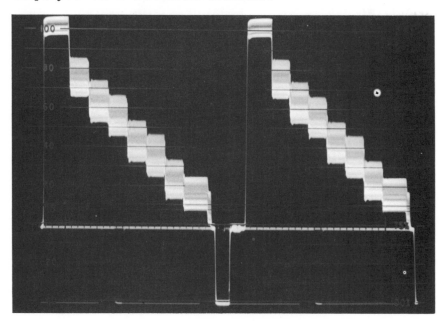

Figure 4.38
Pedestal Adjustments

VIDEO GAIN

Once the PEDESTAL has been established at the proper level, the VIDEO GAIN control is adjusted so that the luminance portion of the video signal never exceeds 100 IRE units. (On some processors, PEDESTAL and VIDEO GAIN interact.) The VIDEO GAIN control effectively adjusts the picture CONTRAST by varying the peak-to-peak voltage amplitude. With PEDESTAL and VIDEO GAIN properly adjusted, video luminance signals should never be less than 7.5 units nor more than 100 units above the blanking level. (Color subcarrier signals may fall to -20 IRE.)

CHROMA GAIN

As shown in Figure 4.39, the CHROMA GAIN control adjusts the amplitude of the color subcarrier (and the picture's color saturation). In use, it is adjusted for the desired color saturation of a scene. Proper adjustment of CHROMA GAIN requires a waveform monitor or vectorscope and a color test signal with a known amount of subcarrier (like color bars or modulated stairstep) to reestablish the original peak-to-peak amplitude of the color subcarrier. Adjustment of the CHROMA GAIN control changes the relative chrominance-to-luminance gain encountered by the video signal. (On some units, CHROMA GAIN and VIDEO GAIN interact.)

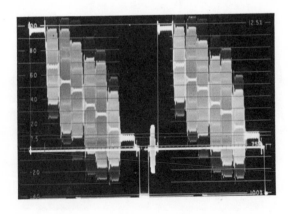

Figure 4.39
Chroma Gain Adjustments

BURST PHASE

The BURST PHASE control on a video processing amplifier varies the phase of the color burst reference signal relative to the chroma subcarrier information. BURST PHASE adjusts for proper hue of the picture on a video monitor that is displaying the signal. As shown in Figure 4.40, BURST PHASE is adjusted so that the color bar dots fall in the appropriate graticule boxes on the vectorscope while the burst reference remains in the nine o'clock position.

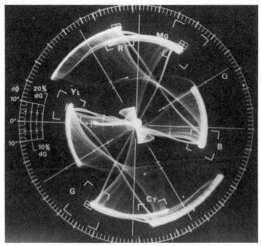

Figure 4.40
Burst Phase Adjustments

BURST AMPLITUDE

As shown in Figure 4.41, the BURST AMPLITUDE or BURST GAIN control adjusts the peak-to-peak voltage of the burst reference signal to reduce phase (hue) errors in monitoring equipment. The burst should be adjusted to peak at ±20 units around the blanking level on a waveform monitor or to the proper burst mark on a vectorscope's graticule, regardless of picture content. (On some equipment, the BURST AMPLITUDE, CHROMA AMPLITUDE, and VIDEO AMPLITUDE controls interact.)

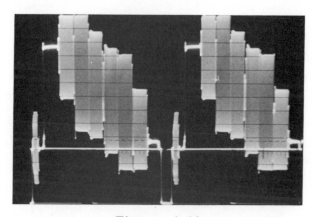

Figure 4.41
Burst Amplitude Adjustments

SYNC AMPLITUDE

As shown in Figure 4.42, the amplitude of the sync signal from blanking level to the sync pulse tip should be 40 units on a waveform monitor, regardless of picture content. If the sync signals are not strong enough, display and recording equipment cannot properly lock to the video signal. If the sync signals are too strong, they may overdrive the display and recording equipment to the point of producing visible distortions in the picture. (On some equipment, SYNC AMPLITUDE and VIDEO AMPLITUDE interact.)

Proper adjustment of a Proc Amp requires a way to measure the various voltages and phases in a composite video signal. Both a waveform monitor and a vectorscope should always be used for a complete and accurate set-up.

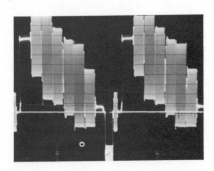

Figure 4.42
Sync Amplitude Adjustments

CHROMINANCE-TO-LUMINANCE DELAY

There is one operator control that is often found on TBCs but rarely on Proc Amps to adjust for any relative delay between the chrominance (picture color information) and luminance (picture brightness information) portions of the video signal. The "C/L DELAY" (chrominance-to-luminance delay) adjustment corrects any relative timing error. These errors are frequently encountered in VTRs because of the separate processing of chrominance and luminance for color correction. Proper adjustment of the C/L DELAY control can be accomplished by using a signal that has exactly the same timing of chrominance and luminance (such as the white vertical stripe on a solid color background shown in Figure 4.43). When viewed on a waveform monitor, the transition points between dark and light of the luminance signal should occur at the same time as the start and stop of subcarrier information.

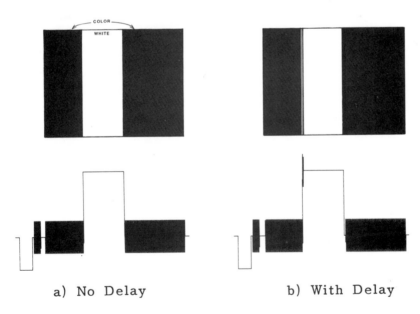

a) No Delay   b) With Delay

Figure 4.43
Chrominance-to-Luminance Delay Adjustment

A special video test signal has been developed that consists of a subcarrier-frequency signal with amplitude variations that fit within an "envelope" that has a frequency approximating the luminance signal of a typical scene. As shown in Figure 4.44, if this "12.5T modulated" test signal is undistorted, the baseline under the pulse appears flat on a waveform monitor. If the there is a bend in the baseline centered under the luminance envelope, there is more or less chrominance gain than luminance gain (C/L gain inequality). If the bend or the baseline crossing point of the bend is not horizontally centered under the luminance envelope, there is relative delay between the luminance and chrominance signals.

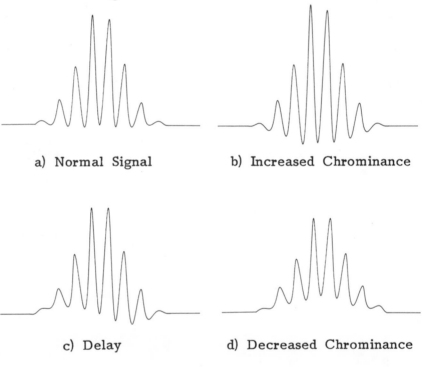

a) Normal Signal   b) Increased Chrominance

c) Delay   d) Decreased Chrominance

Figure 4.44
Modulated 12.5T Signal

COLOR CORRECTION

Once the red, green, and blue video signals have been encoded into the NTSC color signal, it becomes difficult to correct for any white balance or black balance errors. Adjustment of the BURST PHASE control on a processing amplifier or time base corrector acts just like adjustment of the HUE control on a color picture monitor - all the colors are adjusted equally. Remembering that the color problems introduced in white and black balance are the result of *relative* gain inequalities among the red, green, and blue signals, equal adjustment of all colors does not correct the error.

Figure 4.45 shows one method of correcting for white and black balance errors once the signal has been encoded. This method requires decoding the signal into its original red, green, and blue components; then readjusting the gain of the red, green, and blue signals; then reencoding the signal back into an encoded color signal. This method of color correction does work but introduces a considerable amount of noise and distortion into the final signal.

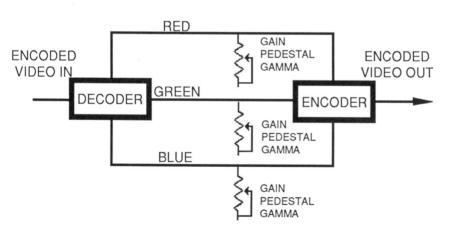

Figure 4.45
Decoding to RGB for Color Correction

Another method of correcting for encoded signal color balance errors adds a complete frame of correcting color signal, as shown in Figure 4.46. This new color information can come from the background generator in a switcher or may be generated by a "color corrector" similar to that shown in Figure 4.47. (The color background generator in a switcher corrects equally for black, gray and white balance inequality whether or not they are all equally out of adjustment.) In essence, the gain imbalance of the original signals is corrected by adding a small amount of the deficient color (which is the "complement" of the predominant color in the picture). Adding additional chrominance to the original signal may necessitate readjustment of the pedestal or gain of the video signal. This color correction method is preferable to the decode/re-encode technique because it introduces less noise and distortion. It's still not as good as getting the color balance correct before the signal is encoded.

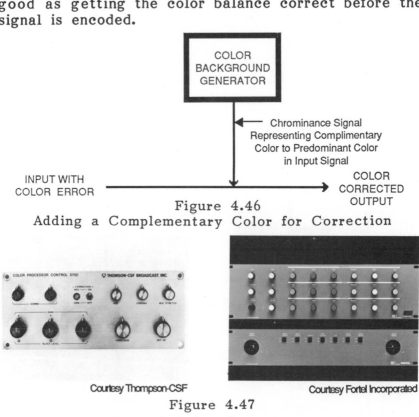

Figure 4.46
Adding a Complementary Color for Correction

Courtesy Thompson-CSF   Courtesy Fortel Incorporated

Figure 4.47
Typical Color Correctors

GAMMA

Gamma describes the accuracy of the light intensity transfer through the television system. Unless the television system is properly adjusted, an increase in light that is sensed by the camera may not be equal to a corresponding increase in video monitor light output. If a graph of the light input versus the light output is plotted as shown in Figure 4.48, the slope (y-axis change versus x-axis change) of the response line of the television system is called "gamma." For a perfect system, where light input changes equal light output changes, the total system gamma equals 1.0.

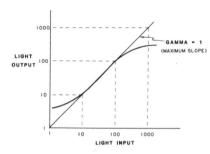

Figure 4.48
Gamma

Various components of the system simply cannot be manufactured to produce a linear light transfer. As shown in Figure 4.49, vidicon pick-up tubes usually have a gamma of 0.65. This, in effect, would produce a picture with only 65% of the contrast of the real scene.

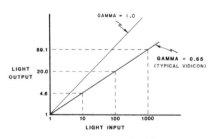

Figure 4.49
Vidicon Gamma

253

Semiconductor pick-up tubes (Plumbicon®, Saticon®, etc.) and solid-state pick-up devices usually have a gamma of 1.0 to produce a realistic contrast rendition of the original scene.

As shown in Figure 4.50, picture tubes in video monitors have a gamma of 2.2, producing a much higher brightness contrast than the original scene. In fact, if a video signal with accurate gain/brightness characteristics (gamma=1) were input directly to a picture tube, the displayed scene would have 220% the contrast of the original scene.

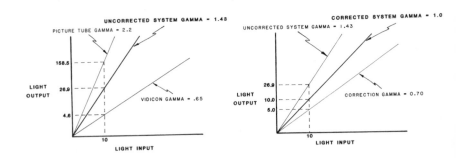

Figure 4.50
Gamma Correction

To ensure faithful reproduction in a television system using vidicons and picture tubes, a correction factor must be derived. Uncorrected, the system has a gamma of 0.65 X 2.2 = 1.43. The circuits between the vidicon and picture tube must have a gamma correction factor of 1 / 1.43 = .70 to produce a gamma of 1.43 X 0.70 = 1.0.

In multi-device color television cameras, gamma controls are frequently used to assure color balance in the middle video signal voltages (where the various shades of gray should appear). As shown in Figure 4.51, the red, green and blue GAIN controls are used when automatically or manually adjusting "white balance" in a camera. Red, green and blue PEDESTAL controls are used when automatically or manually adjusting "black balance." Red, green and blue GAMMA controls are used when manually adjusting "gray balance."

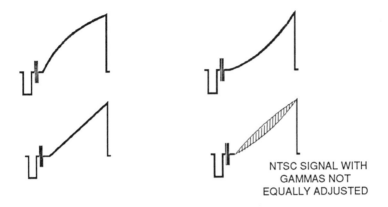

NTSC SIGNAL WITH
GAMMAS NOT
EQUALLY ADJUSTED

Figure 4.51
"Gray Balance" Adjustment of Gamma
in a Multiple-Device Color Camera

Many cameras are equipped with BLACK STRETCH or WHITE STRETCH controls to change the light response in the black or white picture areas. These controls effectively change the gamma in selected contrast ranges. BLACK STRETCH may be used to enhance picture details in dark areas of the scene without increasing the contrast among picture details in light areas. WHITE STRETCH changes the contrast ratio in light scene areas while leaving dark scene areas alone.

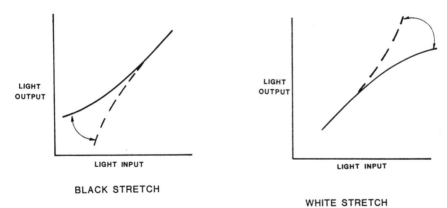

LIGHT OUTPUT

LIGHT INPUT

BLACK STRETCH

LIGHT OUTPUT

LIGHT INPUT

WHITE STRETCH

Figure 4.52
Black Stretch and White Stretch

QUESTIONS

1. Can a color sync generator be locked to a monochrome sync generator?

2. If it were visible in the picture, what hue would burst be?

3. Where is luminance displayed on a vectorscope? White? Black?

4. Can luminance level go below blanking without creating a problem using the signal?

5. Can chrominance go below blanking without creating a problem using the signal?

6. What is pedestal and why is it there?

7. Luminance information is limited to what range of IRE units and what range of voltage?

8. Can a general-purpose oscilloscope be used as a waveform monitor?

9. What one signal parameter can be measured on a vectorscope that cannot be measured on a waveform monitor?

CHAPTER 5

TELEVISION SYSTEMS

TV SYSTEMS

Television production frequently requires the integration of several cameras, monitors, recorders, and test equipment into a comprehensive system. To efficiently combine the parts into a complete television system, additional equipment is frequently required for switching, effects, character generation, distribution, and transfer of motion pictures into video. Getting all this equipment to properly interact presents new and complex problems concerned with the timing of the video signals and their associated sync signals.

VIDEO PRODUCTION SWITCHERS

Although originally designed to simply switch from one video signal source to another, video production switchers have evolved into sophisticated pieces of equipment that allow complex transitions among two or more sources. Figures 5.1 and 5.2 show how complex some of these units can be. The instantaneous transition, or "cut," is the most frequently used way of changing from one video signal source to another, but many refinements to video signal transitions have been developed to allow production flexibility and creativity.

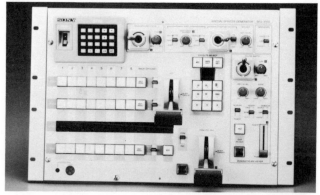

Courtesy of Sony Corporation

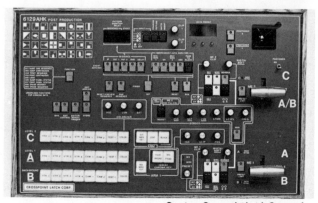

Courtesy Crosspoint Latch Corporation

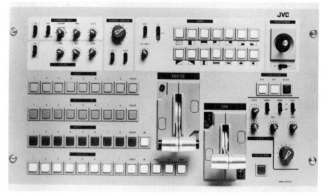

Courtesy JVC Professional Products Company

Figure 5.1
Typical Video Production Switchers

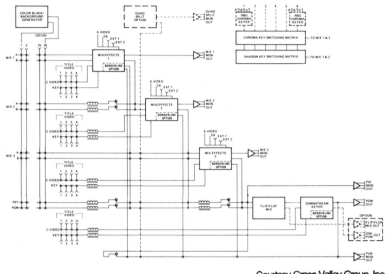

Courtesy Grass Valley Group, Inc.

Figure 5.2
Typical Video Switcher Schematic

VERTICAL INTERVAL SWITCHING

Originally, video production switchers were simply pieces of conducting material which, when a button was pressed, closed the contact to provide an electrical path for the video signal from input to output. As shown in Figure 5.3, whenever contact was not maintained to the output, the video signal was connected to a 75 Ohm resistor to maintain impedance match and reduce system distortions.

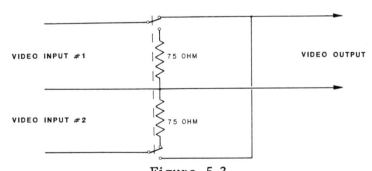

Figure 5.3
Terminating Unused Video Switcher Inputs

But it soon became common practice to defer execution of a cut until the next vertical sync pulse arrives. If the two video signals to be switched are synchronous, the vertical sync pulses of each source should arrive at the same time and switch the signal while the video signal is blanked for retrace from bottom-to-top of the picture. Such "vertical interval switching" eliminates the visible division of a scanned field between picture information from the two video signal sources (because the transition does not occur in the visible picture area). Such a picture often consists of one-half a field of one picture and one-half a field of the other picture. The split picture shown in Figure 5.4, although visible only for a fraction of a second, can be disconcerting to the viewer.

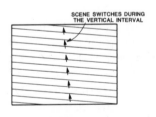

SCENE SWITCHES DURING
THE VERTICAL INTERVAL

 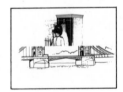 

Figure 5.4
Vertical Interval Switching

As shown in Figure 5.5, vertical interval switching uses an electrical "AND" circuit to command a switch of the path of the video signal to the output, only when both the push-button switch and the vertical sync pulse are received. When another input is selected, the video signal previously selected must be reset to "off" for proper termination (In many electronic switchers, the video signal is terminated at the input and isolation circuits are provided to ensure proper operation without distortions.)

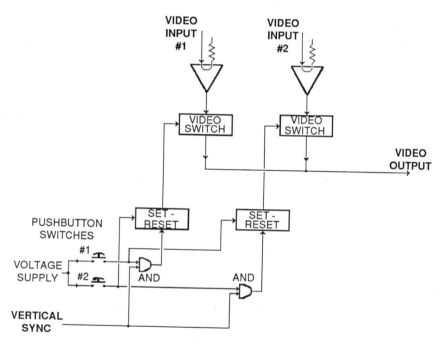

Figure 5.5
An AND Circuit for Vertical Interval Switching

VIDEO SWITCHER BUSES

The video switches themselves are usually grouped into "buses" that allow the operator to connect the one desired input video signal to the output. The buses are used as building blocks to select the video signals to be used in mix and effects operations, as well as preview and program video output signals. For ease in understanding, switcher bus schematics are usually drawn as shown in Figure 5.6.

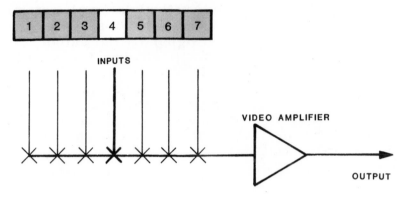

Figure 5.6
Video Switcher Bus

Some switchers provide a "cut bar" that allows one signal to be connected to the program output (on the air) and another signal to be connected to a "preset" bus. As shown in Figure 5.7, pressing the cut bar interchanges the sources to program and preset buses.

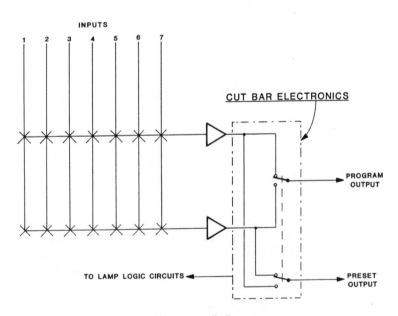

Figure 5.7
Video Switcher Cut Bar

AUXILIARY SIGNAL SWITCHING

Because of the vertical interval switching techniques used for production, provision is sometimes made in the video production switcher to allow a non-synchronous source to be switched to the output using conventional switching techniques (and cause the video monitors to roll). Normally occurring downstream (electronically past) the mix and effects buses as shown in Figure 5.8, these "auxiliary inputs" allow non-synchronous signals from the outside world (like a network feed, remote camera, other stations, etc.) to be used in production with a minimum of operational problems.

In many video switchers, non-synchronous video signals may be sent to normal video inputs to allow cuts-only operation (no special effects). These switchers often have some form of warning to indicate non-synchronous signals.

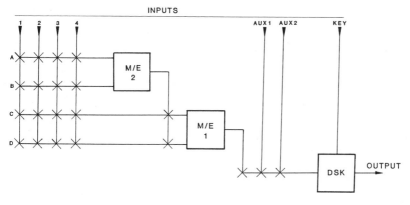

Figure 5.8
Video Switcher Auxiliary Inputs

MIX TRANSITIONS

Creative demands have dictated development of many other types of video signal transitions. The second most common transition slowly "dissolves" or "fades" from one video signal to another. A "fade" is a dissolve to or from a black video signal. During these "mix transitions," both video signals are combined in ratios that are varied by an operator moving the "fader bars" on the control panel of a video switcher. The switcher's video signal output during a dissolve never varies from the standard 1 Volt peak-to-peak, dictating careful control of the gain of each of the video signals. As shown in Figure 5.9, the gains of the output amplifiers in two switch buses are determined by the fader bar controls. With a mechanical linkage between the amplifier gain controls and careful matching of the response of the two video signal amplifiers, the two output signals are simply added together to form the video output signal. When the dissolve is stopped halfway through the transition (to produce a superimposition, or "super"), each of the video signal amplifiers is allowing only half of the video signal to pass. When combined, the voltage of each half is added to give the whole 1-Volt output video signal.

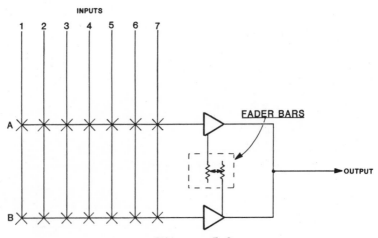

Figure 5.9
Switcher Fader Bars
Control the Gain of Two Amplifiers

SPECIAL EFFECTS

Transitions other than the cut or mix transitions are considered special effects. A "split-screen" picture that is the result of deliberately viewable transitions among two or more video signals is the most often used special effect. If a split-screen picture has a vertical transition as shown in Figure 5.10, *each horizontal scan* in the picture is rapidly switched between the two sources. If the picture has a horizontal transition (with Picture C on the top and Picture D on the bottom); each of the horizontal scans above the transition contains the corresponding picture information from Picture C, and each of the horizontal scans below the transition contains corresponding information from Picture D.

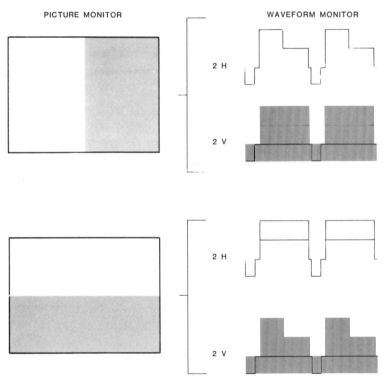

Figure 5.10
Wipe Effects

Split-screen effects other than straight line transitions (circles, curves, etc.) simply alter the time relative to the sync pulses (and the relative position in the picture) that the vertical and horizontal transition points are commanded to occur. As shown in Figure 5.11, many complex split-screen effects may require two or more transition points on an individual horizontal or vertical scan to complete the desired shape of the split-screen insert. These transition times may be altered by adjusting controls on the switcher's control panel to determine the size, shape, and symmetry of the picture's insert shape.

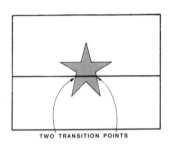

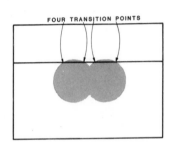

Figure 5.11
Complex Wipe Effects May
Have Several Transition Points

A "wipe" moves the transition points of the split-screen effects across the screen as commanded by settings of controls on the switcher's control panel. These controls are often the same fader bars that are used in the mix transitions, but their function is changed to provide control for varying the time delay commanded to the switching circuits.

On many switchers, provision is made for NORMAL, NORMAL/REVERSE and REVERSE wipe operating modes. As shown in Figure 5.12, when in the NORMAL/REVERSE mode, a wipe goes across the screen in one direction when the fader bars are moved in one direction, and the wipe reverses direction when fader bar motion is reversed. In the NORMAL or in the REVERSE mode, the wipe retains the same direction regardless of the direction of motion of the controls. (There are many, many variations on the wipe direction nomenclature and the instruction manual for an individual switcher model should be consulted when needed.)

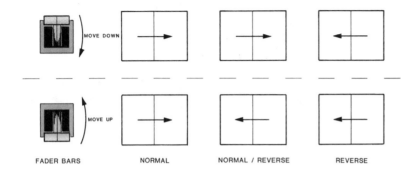

Figure 5.12
Wipe Directions

Many switchers offer the ability to "wiggle" the outline of the insert scene by rapidly changing the position of the transition points between the video signals. As shown in Figure 5.13, this is accomplished by "modulating" the timing of the transition points with a third signal.

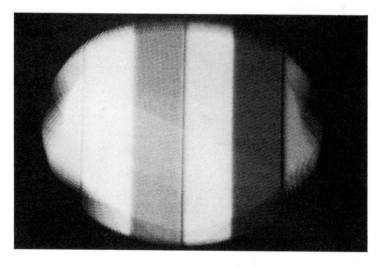

Figure 5.13
Modulated Wipe Pattern

271

A recently developed wipe technique looks more like a rough dissolve instead of the sharp wipe transition with which we are familiar. This "scan wipe" mode randomly selects horizontal scans or random-length portions of horizontal scans to switch from one source to another. For example, if Camera 1 is active on the scene and we are starting a scan wipe, more and more horizontal scans from Camera 2 are switched into the scene. The number of horizontal scans from Camera 2 eventually takes over the entire scene as the scan wipe progresses.

KEY EFFECTS

A real scene may be used to determine the shape of a split-screen effect. Using a "key" camera to generate a video signal that carries picture information about the shape of the desired insert, a voltage transition above or below a "threshold voltage" (or brightness) establishes the position of the transition points. As shown in Figure 5.14, the THRESHOLD control on the switcher's control panel sets the luminance voltage that is used to determine the transition points and is adjusted to reduce the effects of noise and confusion about the brightness voltage of picture details. The two video signals involved in the finished picture can include the key camera and another source, or two independent video signals that use the key camera video to determine the transition.

It is often desirable to put titles or characters on any or all the video signals that are output from a production switcher. In such cases, a "downstream keyer" (DSK) is placed immediately before the signal is output for keys into *any* operating mode of the switcher at the operator's option.

Some sophisticated video production switchers have a circuit that uses varying hue (instead of luminance) from a key camera to determine the special effects transition points. Using this "chroma key" capability, the operator adjusts the HUE control to define the phase of the desired key signal and a THRESHOLD control to define the minimum color saturation to define the special effects transition points, as shown in Figure 5.15.

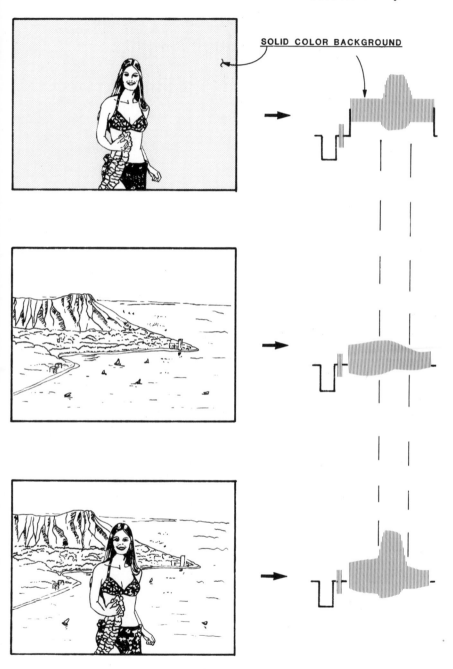

Figure 5.14
Key Concepts

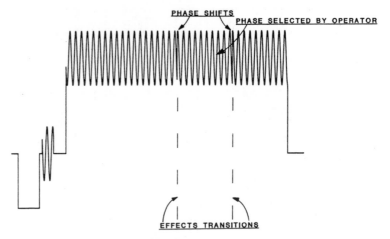

Figure 5.15
Chroma Key Concepts

Chroma key is frequently used in television production when talent in a studio is to appear in front of a background different than the one available. Figure 5.16, shows how the talent is placed in front of a solid color background and the video signal output of the key camera is examined for a specific hue content. When the chroma key switcher detects the pre-selected video signal phase (hue) of the backdrop color, the output video signal is automatically switched to the artificial background picture information. When little or no background color is detected (as determined by the HUE and CLIP or THRESHOLD controls), the output of the key camera is selected.

Figure 5.16
Typical Chroma Key Usage

As shown in Figure 5.17, chroma key units are manufactured to control the picture transitions defined by separate red, green, and blue video signals (R-G-B) as fed to an encoder input or by using encoded (NTSC, PAL, etc.) video signals. In general, the transition points used by chroma key generators that use encoded video signals are not as clean and sharp as an R-G-B unit because of the video noise and signal distortions routinely encountered in the color signal encoding and decoding processes.

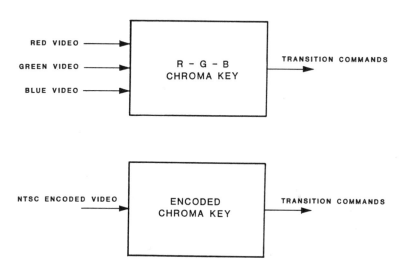

RED VIDEO ⟶

GREEN VIDEO ⟶ R – G – B
CHROMA KEY   TRANSITION COMMANDS ⟶

BLUE VIDEO ⟶

NTSC ENCODED VIDEO ⟶ ENCODED
CHROMA KEY   TRANSITION COMMANDS ⟶

Figure 5.17
R-G-B and Encoded Chroma Key Circuits

Lighting the scene for chroma key is the most critical consideration for successful usage. If the light does not produce a clear delineation between the hues as viewed by the television camera, the transition points are not clearly defined and a ragged edge around the key subject is evident. The lighting can improve the quality of the transitions if foreground objects are rimmed with light of the complementary color of the background. For example, a yellow (red+green) backlight improves the quality of a chroma key based on a blue background.

The choice of hue on which to key are also an important factor in generating a clean chroma key. On many sets, a "chroma key blue" or green background is used for chroma key because there is less blue and green than any other hue in skin tones.

Key and chroma key effects may use a solid color "matte" as a video signal to color the characters in titles or symbols. By using matte and key effects, colored titles and symbols can be placed in a picture without the background scene showing through (as would happen in mix transitions). The hue, saturation, and brightness of the inserted matte is operator adjustable to any desired color.

The matte capability can also be used to create a shadow or outline of a key. As shown in Figure 5.18, a shadow can be created on the characters from a character generator by delaying the characters as they would be cut out of the matte. The delay applied to the matte characters shifts the characters horizontally or vertically, depending on the amount of delay. The delayed matte characters are then keyed over the background picture and the characters are then keyed over both to produce a "drop shadow."

SWITCHER REENTRY

To allow easy operation and creative freedom, many buses are found in a typical video production switcher. As shown in Figure 5.19, two buses are frequently interconnected through one mix/effects electronics module, the output of which is available as an input to yet another mix/effects module. Such "reentry" techniques allow effects on effects, mix or fade to effects, and other sophisticated production techniques.

Adding only a few features to a video production switcher can increase the cost exponentially because of the timing requirements among the myriad possible operating configurations that must be addressed to maintain the sync and subcarrier timing within strict standards. There are many, many variations and options in video production switchers that must be carefully weighed between production capability and cost when considering purchase.

KEY SOURCE

SOLID COLOR MATTE
(FOR CHARACTER)

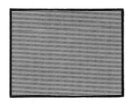

SOLID COLOR MATTE
(FOR SHADOW)

SHADOW COLOR
DELAYED RELATIVE TO
CHARACTER COLOR

Figure 5.18
Drop Shadow Concepts

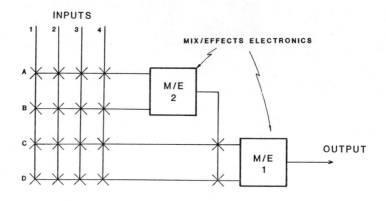

a) Single Reentry

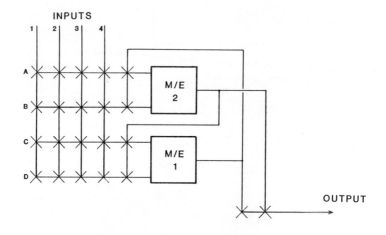

b) Double Reentry

Figure 5.19
Typical Reentry Switcher Schematic

DIGITAL VIDEO EFFECTS

But even these special effects have not been sufficient to fulfill the desire for new and different transitions. Digital video effects (DVE) have been developed to squeeze, enlarge, and relocate entire frames within the picture; or even to rotate pictures. DVE operates on the same principles as a time base corrector (as was discussed in Chapter 4). The incoming video signal is temporarily placed in an array of storage devices. In the frame synchronizer, as the stored video signal is needed, it is read out of memory in exactly the same sequence as stored. But in frame synchronizers controlled by DVE units, the picture information is read out of memory in an order commanded by a microcomputer. As shown in Figure 5.20, the operator has controls to set up the microcomputer to generate the desired effect.

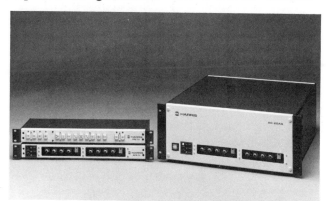

Courtesy Harris Corporation

Courtesy Ampex Corporation

Figure 5.20
Typical DVE Units

For example, if the DVE operator sets up the controls for a reversed and inverted picture, the computer interprets the operator commands to control the sequence of memory read-out of the stored picture information in exactly the reverse order with which it was recorded. When this is displayed on a video monitor (that still scans in the normal way), the resultant picture is inverted and reversed.

With the computer control, DVE can also "zoom in" on certain picture areas by looking at the picture information in only a designated area of the storage array or "expand" the entire picture by automatically filling in any missing video information from the surrounding picture areas. Figure 5.21 shows a DVE control panel.

All the DVE processes can be dynamically controlled by the computer to generate moving reduced frames across a picture, a series of stored "snap shots" of past frames, rotating frames, or almost any other conceivable effect. The output of the DVE is then available as another video source for use with any of the other transition modes of a video production switcher.

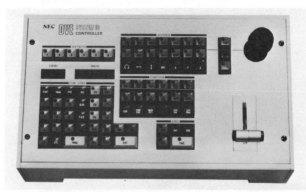

Courtesy NEC America, Inc.
Broadcast Equipment Division

Figure 5.21
Typical DVE Control Panel

SYSTEM TIMING

In television systems with several sources of video signals, *each of the video signals must be timed so that each signal's horizontal and vertical sync pulses arrive at the input to the video production switcher simultaneously.* Delay may be caused by video signal cable runs, video signal delays within the source equipment, sync signal delays, or sync signal cable runs.

When timing a system, one first establishes the point where timing is critical. This "zero time point" is usually at the input connectors of the video switcher. Video production switcher manufacturers design their equipment to equalize the delay inequalities that may occur among the various buses and electronics within the switcher, allowing the system to be timed to the switcher inputs. Then, the system is examined for the delay encountered by each of the video signal sources.

As shown in Figure 5.22, in a genlock system all horizontal sync pulses must be adjusted by using an H PHASE control on the video source or by using a "delay line" in the sync signal path on the way to the video source. Likewise, the phase of the subcarrier reference bursts of all the video signals that are input to the switcher can be adjusted to be in phase by using the SUBCARRIER PHASE control at the video source or by using delay lines in the subcarrier signal paths to the video sources.

A system timing error is particularly noticeable when performing a mix transition on a video production switcher. In this mode, a video switcher is slowly changing its output video signal between two incoming composite video signals, but retains the same sync and burst references until the transition is almost complete. As shown in Figure 5.23, when the sync and burst are instantly switched (toward the end of the transition), the picture jumps in position or hue.

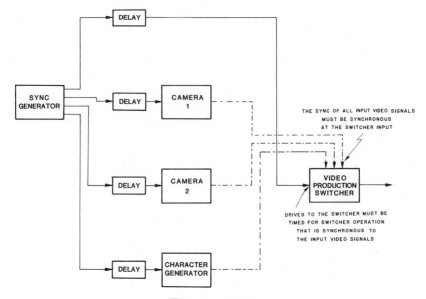

Figure 5.22
Genlock System Timing Concepts

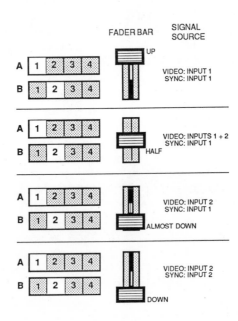

Figure 5.23
Switching Between Two Signals

As an example, select Camera 1 on the A bus, Camera 2 on the B bus, and move the MIX/EFFECTS (M/E) BARS to the up position. The video and the sync reference signals are from Camera 1. As the M/E BARS are slowly lowered to the down position, the video source changes from Camera 1 to Camera 2 *while the sync reference signal remains with Camera 1.* When the M/E BARS reach the 90% down position, a switch contact closes to instantaneously change the sync and burst reference from Camera 1 to Camera 2 (and switch the tally lights). As shown in Figure 5.24, if the two signals are not in "H-Phase," a noticeable horizontal jump occurs in the Camera 2 picture when the switch changes sync reference. For the same reason, a difference in burst phase can produce an equally objectionable color shift.

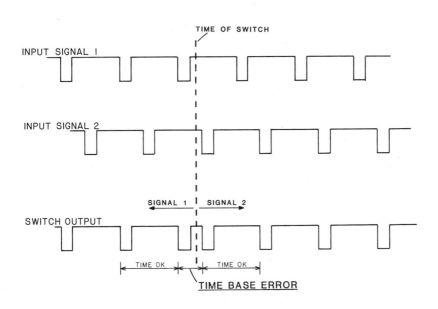

Figure 5.24
Sync Pulses when Switching Between Two Signals

CABLE DELAY

System timing arises from the amount of time it takes for a signal to travel down cables of varying lengths. Because the entire scanning system in television is based on strict timing standards, the various cable lengths in a typical television system introduce various amounts of delay.

The amount of delay in a given length of cable can be calculated by using the formulas:

$$\text{Cable Delay (ns)} = \frac{\text{Cable Length (in)}}{\text{Velocity of Propagation X 11.8}}$$

$$1° \text{ of Subcarrier} = 0.77 \text{ nanoseconds}$$

$$1 \text{ ns} = 1 \text{ nanosecond} = 0.000000001 \text{ second}$$

The "velocity of propagation" is the speed of the electricity as it flows down the cable (expressed as a factor of the velocity of light) and is listed on the cable manufacturer's specification sheet. For Belden 8281, 8279, and 9221, the velocity of propagation is 0.66. For RG-59/U, the velocity of propagation is 0.78. Thus, a video signal traversing from one end of a 100-foot (1200-inch) long piece of RG-59/U cable takes

$$\text{Cable Delay} = \frac{1200}{0.78 \text{ X } 11.8}$$

$$= 130.4 \text{ ns}$$

H PHASE ADJUSTMENT

When timing a television system, the video signal that has encountered the longest time delay is used as the reference timing signal (because the other signals can be easily delayed for time coincidence). In the example shown in Figure 5.25, the H PHASE control on Camera 2 is adjusted so that its leading edge of horizontal sync occurs at exactly the same time as the horizontal sync pulses of Camera 1 *as measured at the input to the switcher.* (If a switcher has a "blanking processor" circuit, it should be switched OFF when adjusting H Phase, then turned back ON.)

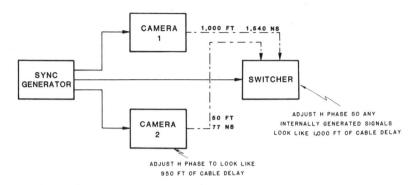

Figure 5.25
H Phase Adjustment Concepts

As shown in Figure 5.26, horizontal phase between two video or sync signals may be measured by using a waveform monitor or a picture monitor with an external sync input. The left vertical edge of the horizontal sync pulses as shown by a waveform monitor should not move horizontally when switching between two signals if the signals are in time. The picture monitor shows no movement of the left edge of horizontal sync (the super-black vertical bar in a cross-pulse display) if the signals are in time. If the monitor has an "A minus B" display mode, the left edge of horizontal sync of the two signals have no white or black outline when the two signals are in time.

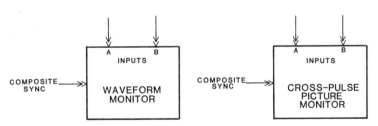

NOTE: WAVEFORM MONITOR OR PICTURE MONITOR MUST USE EXTERNAL SYNC

Figure 5.26
Connecting a Monitor to Display H Phase Adjustments

SC PHASE ADJUSTMENT

Once horizontal timing is adjusted, the phase of the bursts of all the input video signals must be timed together to eliminate color shifts. A vectorscope is used to measure these minute changes in phase. The vectorscope is locked to one video signal (or an external subcarrier reference source) and set to display the other video signal. When the displayed burst lies at the 9 o'clock position the two signals have the same phase.

If the signal is measured after new burst has been inserted (as frequently is done in production switcher), SUBCARRIER PHASE or BURST PHASE may be confirmed by placement of the dots from the source's color bar signal in the vectorscope graticule boxes as shown in Figure 5.27. The reinserted burst then serves as the reference for the color bars. Caution should be exercised if color bars are used to time the H Phase of a signal while viewing a picture monitor. The width of the color bars is adjustable in most generators and a change in color bar width can be easily misinterpreted as a change in H Phase.

In cases where coarse adjustment of timing is required, H and subcarrier phase can be adjusted for no picture position shift or color shift while operating the switcher's fader bars.

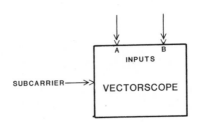

Figure 5.27
Connecting a Vectorscope
to Display S/C Phase Adjustments

SYNC LOCK SYSTEM TIMING

If there is no H PHASE control on the video source, the source may be driven by sync that has been delayed to coincide with the sync that has the longest unalterable delay time. In Figure 5.28, the Camera 1 video signal has the longest delay time (1540 nanoseconds). To have all the sync pulses arrive at the switcher at the same instant, we may delay the sync source for Camera 2 by 1463 nanoseconds (1540 ns - 77ns = 1463 ns). *Although Camera 2 is not really timed to Camera 1, when the signals reach the video switcher, they appear to be timed correctly.* The sync applied to the video switcher must also be delayed to correct the timing of internally generated color backgrounds and processing functions.

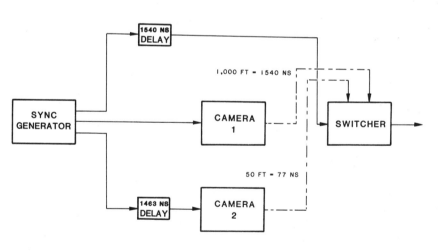

Figure 5.28
Timing Video Signals at the Switcher Input

Because of the delay characteristics of coaxial cable, a cable cut to a calculated length can delay the video or sync reference signal by the desired amount. This method is not as desirable because the coaxial cable introduces more distortion and attenuation than a properly designed and installed delay line. Figure 5.29 shows a commercially available delay line.

A second, less desirable philosophy of timing a system involves delay of the video signals themselves. Because of the distortion produced by placing *any* device in the signal path, this procedure should be used sparingly.

When the sync timing is changed, all the other drive pulses (vertical and horizontal drive, composite blanking, burst flag, etc.) must be delayed to maintain proper timing among the various pulses. This becomes obvious when considering the time interrelationship that must be maintained between sync and blanking signals.

The subcarrier signal may also need to be delayed before use in the video source so that the signals are apparently phased together at the production switcher. Most color origination equipment manufactured today has an internal S/C PHASE or BURST PHASE control. For equipment without such a control, discrete phase shifters, subcarrier distribution amplifiers, and phase shifted sync generator outputs are available.

Figure 5.29
Typical Delay Module

CABLE EQUALIZATION

The construction of a coaxial cable introduces frequency-dependent loss of video signal. As shown in Figure 5.30, as the video signal frequency increases, loss of the video signal increases.

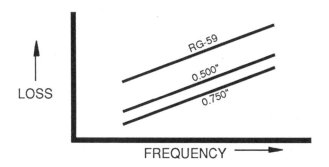

Figure 5.30
A Typical Graph of
Coaxial Cable Loss versus Frequency

As shown in Figure 5.31, a coaxial cable is composed of a center conductor surrounded by an insulating "dielectric" material which is surrounded by a shield and outer insulation. A schematic diagram for a coaxial cable is shown in Figure 5.32 where the center conductor and shields are separated by the insulating dielectric to form a long, distributed capacitor.

Figure 5.31
Coaxial Cable Construction

289

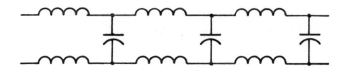

Figure 5.32
Coaxial Cable Schematic Diagram

One electric property of capacitors is called capacitive reactance. Capacitive reactance is the transfer of signal charge from one conductor (the center conductor) to the other (the shield). Coaxial cable losses from capacitive reactance increases and frequency increases. The formula for capacitive reactance is

$$X_C = \frac{1}{2\pi fC}$$

where f=frequency and C=capacitance in farads. X_C is measured in Ohms.

Examination of the capacitive reactance loss formula and the coaxial cable schematic reveal that more and more of the video signal is shorted between the center conductor and the shield as the frequency of that signal increases.

A second loss is caused by inductive reactance. Inductive reactance is the loss induced in a coil of wire (or inductor). Losses from inductive reactance conform to the formula

$$X_L = 2\pi fL$$

where f=frequency and L=inductance (in Henrys). X_L is measured in Ohms.

With two conductors (the center conductor and the shield), a coaxial cable conceptually has two distributed coils of wire with no turns. Although the coils of wire do not have any turns, inductive reactance still comes into play. Examination of the formula and the coaxial cable schematic shows that high frequency components of the video signal is lost much more rapidly than the low frequency components within each conductor of the coaxial cable. As shown in Figure 5.33, as the cable becomes longer, the first portion of the video signal to be lost are the high frequency luminance components. Then the chrominance is lost, then mid-frequency luminance, and finally low-frequency luminance and sync.
ce components. Then the chrominance is lost, then mid-frequency luminance, and finally low-frequency luminance and sync.

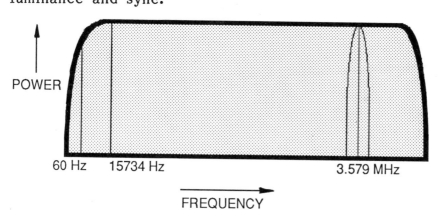

Figure 5.33
Video Signal Frequency Distribution

Capacitive reactance and inductive reactance conspire to place a practical limit on the length of a coaxial cable carrying a video signal. The larger the size of the cable, the longer it may be before the signal significantly deteriorates. A cable one-quarter inch in diameter (like RG-59) can carry a video signal about one hundred feet before seeing significant degradation and about three hundred feet before being totally unusable. A three-quarter inch cable can carry a video signal about three hundred feet before seeing significant degradation.

Cable equalizing amplifiers similar to those shown in Figure 5.34 are available to optimize high-frequency signal transfer. Many Camera Control Units (CCUs) have a built-in cable equalizing circuit for long camera cable runs. Figure 5.35 shows the control used to adjust for cable length on a typical CCU. The CABLE LENGTH control on a CCU does not adjust for delay, just for equalization.

Cable equalization is frequently adjusted using a multiburst signal or a color bar signal. When using a color bar signal, the amplitude level of chrominance is adjusted to match the amplitude level of luminance.

Courtesy Grass Valley Group, Inc.   Courtesy Omicron Video

Figure 5.34
Typical Cable Equalizing Amplifiers

Figure 5.35
CCU Adjustment for Camera Cable Equalization

FRAME SYNCHRONIZERS, FIELD SYNCHRONIZERS, AND TIME BASE CORRECTORS

Recent developments in video signal storage technology has produced equipment that can store complete fields or frames of picture information for still-frame reproduction or remote source synchronization. Figures 5.36 and 5.37 show some of these frame or field "synchronizers" or "stores" (depending on the amount of storage available in the unit). These units can be used to correct extreme time base errors or synchronize two video signals that are non-synchronous. In broadcast applications, these units are often used to incorporate non-synchronous signals from remote news vans into a studio-originated newscast.

Courtesy Harris Corporation

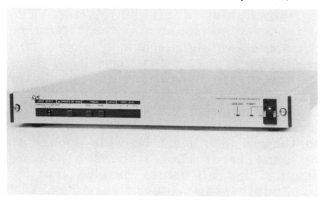

Courtesy JVC Professional Products Company

Figure 5.36
Typical Synchronizers

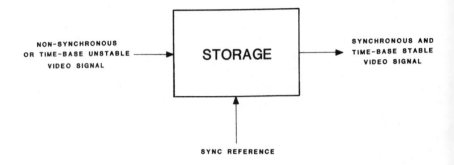

Figure 5.37
Typical Synchronizer Schematic

On some storage units, an additional distinction is made between TBC and synchronizer operation by the amount of sync signal instability that is anticipated. Video tape sources normally have greater signal instability than other sources.

A/B ROLL

Normally, two time base correctors are needed to perform dissolves and special effects between two video tape sources; however, under dire need, an A/B roll may be accomplished using only one TBC. The interconnection shown in Figure 5.38 shows how the one TBC that is correcting only the VTR-B signal may be genlocked to the VTR-A signal. With this mode of operation, the output of the TBC is the VTR-B video signal that is synchronous to the referenced VTR-A signal. With the two signals now synchronous, dissolves and special effects may be performed.

There are severe limitations and precautions when using one TBC for A/B-Roll techniques. First, the time base of the signal from VTR-A must be very stable. Second, the acceptance of the genlock input of the TBC and the switcher must be wide enough to accept the time base error that is present on the VTR-A video signal. Third, all cables must be kept as short as possible. Fourth, the video signal from VTR-A must provide a continuous signal through the entire edit, including all the time that the VTR-B signal is active.

(If the VTR-A signal changes sync timing while the VTR-B signal is active, a glitch in the VTR-B picture appears. The VTR-A signal serves as the reference for the TBC and switcher.)

This same technique of locking to the output of a video tape recorder may be used to lock cameras, character generators, and other potential signal sources for post-production activities.

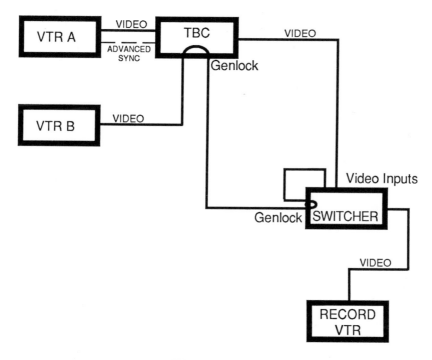

Figure 5.38
Schematic of an A/B Roll Editing System
Using One TBC

CHARACTER AND GRAPHICS GENERATORS

Generation of television signals from equipment other than a camera is a complex marriage of digital computer technology with an analog television world using equipment similar to that shown in Figure 5.39. Initially, character generators were developed simply to put alphanumeric characters (letters and numbers) on the screen; however, the practice of creating drawings and animation with "graphics generators" is rapidly growing.

Courtesy Thompson-CSF

Courtesy JVC Professional Products Company

Courtesy Quanta Corporation

Courtesy Quanta Corporation

Figure 5.39
Typical Character and Graphics Generators

Although there are many techniques used in character generation, all use a common technique of taking an entry from a typewriter-style keyboard and placing that character in a designated location on the screen. As shown in Figure 5.40, the television picture is divided into character-size blocks, the contents of which can be individually addressed and filled by microcomputer.

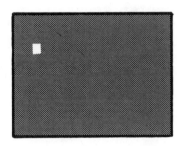

Figure 5.40
A Cursor showing One Character-Size Block

THE CURSOR

A flashing cursor (appearing only on the MONITOR OUTPUT of a character generator) keeps the operator informed of the block location where the next entry can be stored. When a character is entered into the keyboard, a unique digital signal is sent for storage. The storage array remembers picture block location and the character within the block. As each block location is scanned by the television system, the character stored for that block is recalled.

GENERATING CHARACTERS

But in the television system, each block is scanned by many horizontal paths. To provide the proper timing of the information fed to the television system, each computer-generated character is divided into horizontal strips as shown in Figure 5.41. The top character strip is read out at the proper time by using horizontal and vertical sync and the timed location of the information block location as references. The next horizontal sync pulse is used as the timing reference for the read out of strip number 3 (leaving strip number 2 for the interlaced opposite field). The next horizontal sync pulse references the start of the read out of strip number 5, etc.

As shown in Figure 5.42, a microcomputer integrated circuit "chip" installed within the character generator supervises the operations to determine when the information in a block is to be read into or out of memory. A character generator chip determines the sequence of switched picture elements unique to the horizontal strip to be generated, and a timing "latch" determines when (and consequently where in the picture) a character is to be generated.

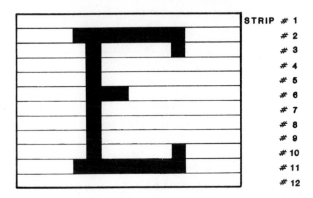

Figure 5.41
Division of a Character into Horizontal Strips

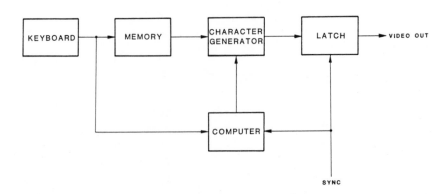

Figure 5.42
Typical Character Generator Block Diagram

298

GENERATING GRAPHICS

Just as a character generator divides a screen into separate blocks, a graphics generator divides the screen into even more (and smaller) information blocks, as shown in Figure 5.43. When a drawing is made, the start and stop of each line (if a straight or geometrically easily derived line) or the series of blocks the line crosses (if drawn freehand) is defined and stored. After the computer interprets the input information, each location in the picture is assigned a digital code that describes the brightness or hue. When displayed on a picture tube, the computer-generated blocks of graphic information merge and appear to form continuous lines or colors.

Figure 5.43
Division of a Picture
Into Small Blocks

The entered lines and points in the picture can also serve as the boundaries for a solid field of color or brightness on proper command to the computer. When this is desired, the computer automatically changes the code defining brightness, hue and saturation of each block within the operator-defined boundary.

I/O DEVICES

To save time and effort, graphics generators often use sophisticated locating equipment to rapidly designate locations in the television raster. Although there are many different techniques in use, most professional graphics equipment uses one of three techniques: a "light pen," "mouse," and "graphics tablet."

A light pen contains a photodiode at the open end of a cylinder to detect the time when the electron beam energizes the phosphors on a television display at the location to which the pen points. As the pen is pointed toward the display, this time information is used in the computer to determine the location of the pen point in the display frame.

A mouse is a small hand-held assembly that is moved across a flat surface. As the mouse moves, sensors roll across the flat surface to maintain position information.

Most popular of the input techniques is the graphics tablet. Using a wand or a mouse-like pick-up device, location in a frame is derived from location in the frame defined by the graphics tablet. Figure 5.44 shows a graphics tablet.

Courtesy Harris Corporation

Figure 5.44
Typical Graphics Tablet

PAGE STORAGE

Even though many units offer enough solid-state memory to retain many pages of generated information, the information as stored in memory is normally lost when the power to the character or graphics generator is turned off, or when a new page of information is entered. Battery power back-up or an outside storage medium similar to the floppy disks shown in Figure 5.45 is required when this information is needed later.

Figure 5.45
Typical Character Generator Storage Media

There are two widespread methods of storage of character or graphics-generated information. Using magnetic storage technology, cassette format audio magnetic tapes and spinning discs with a coating of oxide ("floppy discs") are frequently used for permanent or semipermanent storage of the information. Figure 5.46 shows a typical graphics generator disk storage unit.

Courtesy Chyron Corporation
Figure 5.46
Typical Floppy Disk Storage Unit

FILM CHAIN

It is frequently necessary to convert the picture information from motion picture or slide films into a video signal. One basic technique that may be used in this "film-to-tape transfer" involves a standard television camera that scans a sharply focused image on a screen or lens. A second method of film-to-tape transfer involves moving the film or moving the light source behind the film to create the effect of scanning.

The most obvious method of converting from film to television picture information is to project the film's image onto a screen and focus a normal television camera on the image. This often produces acceptable results if other light is carefully controlled and the camera and projector are controlled to minimize optical distortions like keystoning. (As we shall see later, a special motion picture projector must be used to eliminate the interference between the 24 frames-per-second film rate and the 30 frames-per-second video rate.)

MULTIPLEXERS

In most situations, however, extraneous light may be difficult or impossible to control and other techniques must be used. In conventional "film chains" projectors effectively focus their image in a "field lens" placed between the projector and camera. With a field lens, the light transferred between projector and camera may be carefully controlled and the image quality carefully adjusted. The assembly that mounts the field lens and shades it from extraneous light is called a "uniplexer" (for one projector and one camera) or "multiplexer" (for several projectors or cameras).

Many multiplexers use motorized mirrors to optically switch the desired projector to the television camera. When the Slide Projector is selected in Figure 5.47, no mirror is in the optical path, allowing the image to be projected straight through the field lens to the camera. When Motion Picture Projector 1 is selected, a mirror is placed in the optical path to reflect the projected image through the field lens to the television camera. The Motion Picture Projector 3 path has a second mirror that reflects the image to the camera.

a) Slide Projector Selected

b) Motion Picture Projector #1
Selected

c) Motion Picture Projector #2
Selected

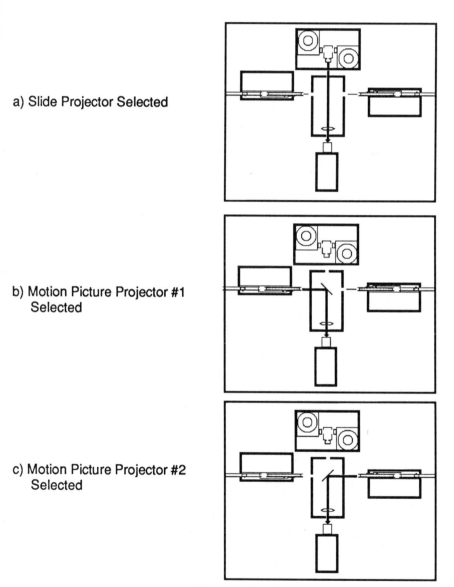

Figure 5.47
Typical Multiplexer Operation

MOTION PICTURE PROJECTORS FOR TV

A motion picture projector must be mechanically adapted to the system because of image interference between the standard 24 frames-per-second film projection rate and the (approximately) 30 frames-per-second television scanning rate. Motion picture projectors normally flash a frame for a brief period then use a rotating shutter to block out the light and shut off the projected image while the film is pulled to the next frame. In projectors not intended for television usage, a "two-blade" shutter similar to that shown in Figure 5.48 is used to project the image while the lamp can shine between the blades at 24 frames-per-second. As shown in Figure 5.49, when the

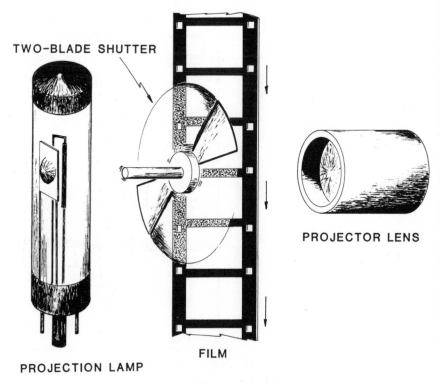

TWO-BLADE SHUTTER

PROJECTOR LENS

FILM

PROJECTION LAMP

Figure 5.48
Two-Blade Shutter

standard 24 frames-per-second film projection rate is projected with a two-blade shutter into a television camera, a horizontal black area occasionally flickers through the television picture. The black bar occurs because the shutter has shut off the light for film pull-down during the visible portion of the television scan.

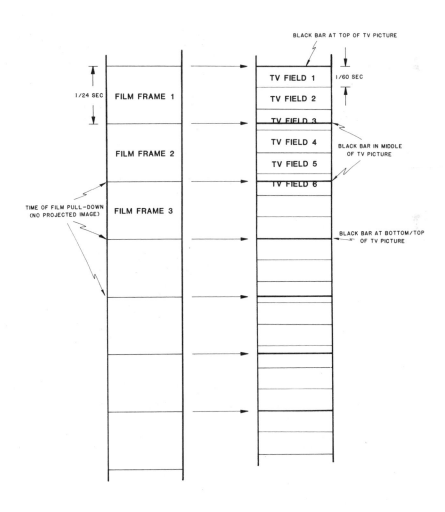

Figure 5.49
Two-Blade Shutter Technique

To eliminate this interference bar, a five-blade shutter similar to that shown in Figure 5.50 is used to project each film frame two or three times before the film is pulled down to project the next frame. Figure 5.51 details how this (2+3)/2 = 2.5 frame projection average produces an effective rate of 2.5 X 24 frames/second = 60 projected film frames per second, the field rate for television. Note that the film is still transported across the projector at the same speed as though it is being shown at 24 frames-per-second making scene action and audio appear normal. With the projector so modified, any visual interference from shutter bars has been reduced below the visually objectionable level.

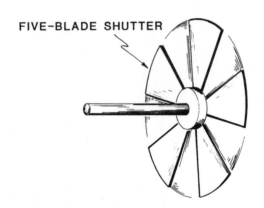

FIVE-BLADE SHUTTER

Figure 5.50
Five-Blade Shutter

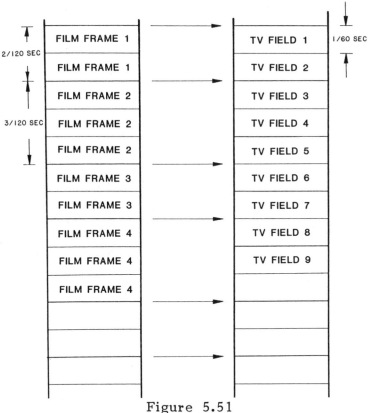

Figure 5.51
Five-Blade Shutter Techniques

CCD TELECINES

Another method of converting motion picture images into video signals pulls the film across a transport similar to that shown in Figure 5.52 without stopping the frame to project the image. The steady motion of the film across the transport is used to create the vertical television scan while a one-line CCD similar to the one shown in Figure 5.53 senses the picture information horizontally. This "CCD Line-Scan" technique takes the image scanned by the CCD sensors (one each for red, green, and blue) stores them in a circuit that stores the picture information from the frame, then reads out the stored information as necessary to effectively create the scan. This storage technique is necessary to create an interlaced readout of the progressive scan from the CCD sensors.

Courtesy BTS Corporation   Courtesy BTS Corporation

Figure 5.52
Typical CCD Telecines

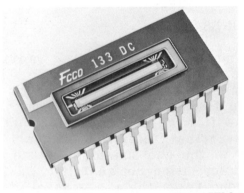

Courtesy BTS Corporation

Figure 5.53
Typical CCD Chip used in CCD Telecines

FLYING SPOT SCANNERS

Another technique, called "flying spot scanning," uses the bright spot that is scanning across a special cathode ray (picture) tube. As shown in Figure 5.54, as the light shines through the film, it creates a scan of the film frame. An inexpensive light sensor that looks at the entire film frame may be used to detect the scanned picture information.

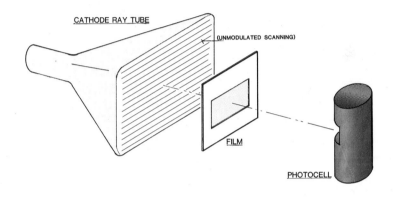

CATHODE RAY TUBE

(UNMODULATED SCANNING)

FILM

PHOTOCELL

Figure 5.54
Flying Spot Scanner Techniques

SIGNAL DISTRIBUTION METHODS

In general, there are two distinct ways of getting the video and audio signals to the end user. "Baseband signals" may be sent down cables as they originally appear at the output of a microphone (audio) or a camera (video). These "raw" video and audio signals are sent down separate cables after being amplified to sufficient power. The second distribution method "modulates" these baseband signals onto radio frequency ("RF") television channels.

BASEBAND SIGNALS

As shown in Figure 5.55, baseband audio may be distributed among various pieces of equipment by paralleling or "bridging" two inputs directly across a single output (provided they are bridging inputs). Note that the problems associated with the matching of impedance, level, and balance configuration remain.

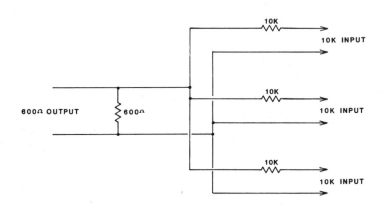

Figure 5.55
Audio Bridging Concepts

Unlike baseband audio, baseband video may not be distributed by bridging the inputs because of much more stringent requirements of matching impedance, level, and balance. This matching is very important because of the considerably higher frequencies within the video signal. To input a single video signal into several pieces of equipment, input circuits are frequently designed to allow a small portion of the signal to be picked off while going through a "loop-through" before sending the remainder of the signal on to downstream equipment. When interconnected as shown in Figure 5.56, several individual pieces of equipment with loop-through inputs may be interconnected to receive the same video signal. Since the loop-through methods are conceptually the same as the audio bridging methods, the number of individual pieces of equipment that can be involved in the circuit is limited. (Audio bridging also has limitations but the lower frequencies in audio signals produce fewer problems.)

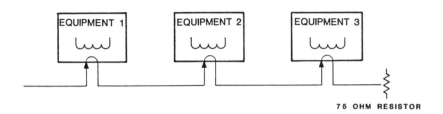

75 OHM RESISTOR

Figure 5.56
Video Loop-Through Concepts

SYSTEM TERMINATIONS

Notice that at the physical end of each cable, after all equipment has been connected in series, a 75 Ohm terminating resistor is used so that the impedance of the incoming signal is matched (and distortion reduced). Many models of equipment have a switch between "HI-Z" and "75" to activate an internal 75 Ohm terminating resistor. Figure 5.57 shows a typical HI-Z/75 switch.

Figure 5.57
Typical HI-Z/75 Switch

DISTRIBUTION AMPLIFIERS

To overcome the problems created by looping several inputs, a "distribution amplifier" (DA) similar to those shown in Figure 5.58 can be installed to terminate the signal at its input and generate several independent output signals. A distribution amplifier consists of two stages, one to split the signal and the other to amplify the signal. A "build-out" circuit takes the one signal and splits it into several independent signals with identical impedance characteristics. As shown in Figure 5.59, an amplifier circuit is designed between the input to the DA module and the input to the signal splitting circuit to overcome the losses associated with the signal splitting.

Courtesy Video Accessory Corp.

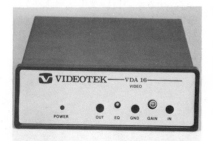

Courtesy Videotek, Inc.

Courtesy Utah Scientific, Inc.

Courtesy Grass Valley Group, Inc.

Figure 5.58
Typical Distribution Amplifiers

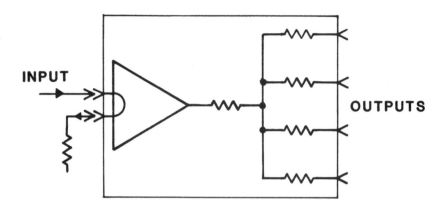

Figure 5.59
Typical Distribution Amplifier Schematic

Each of the DA output signals is isolated from the others, so the effects of impedance mismatch, level inequalities, etc. are minimized. *Although some isolation is provided by a DA, it is good operating practice to terminate any unused outputs with a 75 Ohm resistor to eliminate any distortion to the used output signals.*

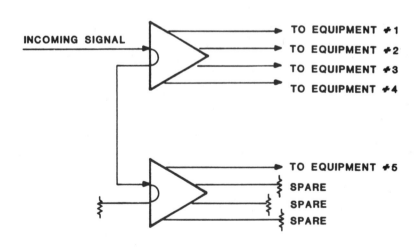

Figure 5.60
Terminating DA Inputs and Outputs

Distribution amplifiers are also available to reduce adverse bridging, impedance, and level effects in audio systems.

In some small baseband distribution systems, both video and audio are carried through a single cable (with several independent conductors) and interconnected between the pieces of equipment with an 8-pin EIAJ (Electronic Industries Association of Japan) connector similar to that shown in Figure 5.61. The outputs and inputs associated with such arrangements are usually high impedance (around 10,000 Ohms) and dictate short cable lengths.

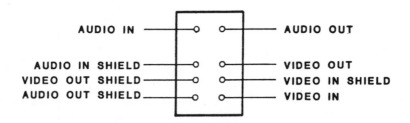

```
AUDIO IN ──────────o   o────── AUDIO OUT

AUDIO IN SHIELD────o   o────── VIDEO OUT
VIDEO OUT SHIELD───o   o────── VIDEO IN SHIELD
AUDIO OUT SHIELD───o   o────── VIDEO IN
```

Figure 5.61
8-pin EIAJ Connector

R.F. SIGNALS

The number of cables for video and audio signals may be minimized by passing a set of video and audio signals through a "modulator" similar to the one shown in Figure 5.62. As shown by the typical TV channel signal shown in Figure 5.63, the modulator acts as a miniature television transmitter to combine the baseband video and audio signals into radio frequency (RF) signals that fall in the frequency band allocated to a broadcast television channel. Figure 5.64 details the modulation procedure. As shown in Figure 5.65, several of these RF signals can be simultaneously transmitted down a cable or through the air (with sufficient power and FCC approval) to the final destination where the signal is demodulated (tuned) back into the original baseband video and audio signals.

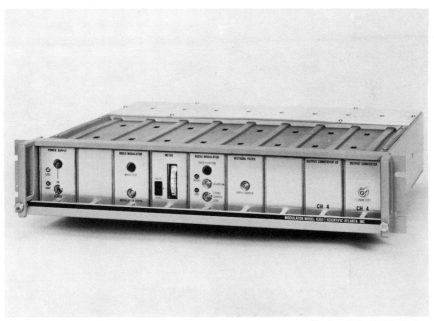

Courtesy Scientific-Atlanta, Inc.

Figure 5.62
Typical TV Modulator

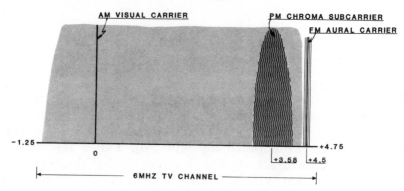

Figure 5.63
Typical TV Channel Spectrum

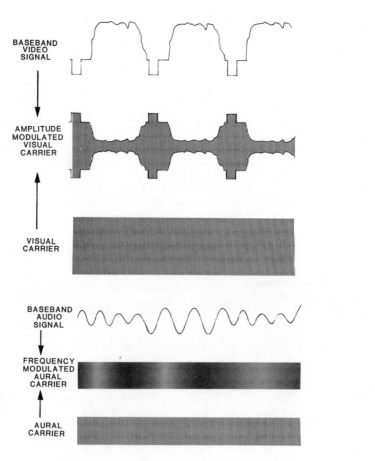

Figure 5.64
RF Signal Modulation Techniques

316

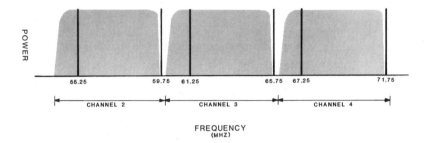

POWER

55.25 59.75 61.25 65.75 67.25 71.75

| CHANNEL 2 | CHANNEL 3 | CHANNEL 4 |

FREQUENCY
(MHZ)

Figure 5.65
RF Signal Spectra on One Cable

Because of the unique properties of high-frequency RF signals, special distribution techniques are required. "Taps" are designed to serve the same purpose as loop-through inputs in baseband video equipment. "Splitters" are used in reverse as "combiners" to bridge two or more equipment outputs together onto the one cable carrying the RF signals.

The primary system design concern when using an RF signal is the power of the signal at the destination. As shown in Figure 5.66, each tap or splitter has an "insertion loss" specification that describes the signal power lost when the device is inserted into the line. When one considers that a splitter simply splits the input signal power into equal parts, if a splitter has two output ports, the ideal insertion loss is:

2-way splitter insertion loss = 10 X log $\frac{1}{2}$ = -3 dB

(This is simply an approximation, for there will always be some additional loss because no device is perfect.)

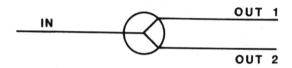

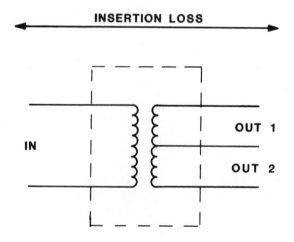

Figure 5.66
Typical Splitter

Taps are available with different insertion loss characteristics to optimize the design of a system. Besides the loss on the main cable, a tap also has a specified "tap loss" to describe the loss that the signal has experienced in getting from the cable input to the tap output of the device.

Special "directional couplers," as shown in Figure 5.67, are available to provide an RF tap with electrical isolation to reduce the downstream effects of any interference that may originate from the tap or from other downstream taps. Besides insertion loss and tap loss specifications, the directional coupler also has a specified "isolation loss" to describe the loss in power that a signal experiences if it tries to go from the tap connector to the output connector (or vice versa).

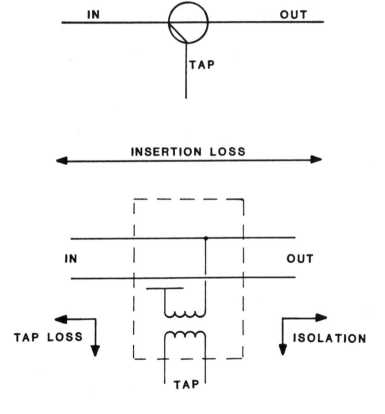

Figure 5.67
Typical Directional Coupler

A system similar to that shown in Figure 5.68 combines splitters and directional couplers to simultaneously distribute RF signals to several destinations.

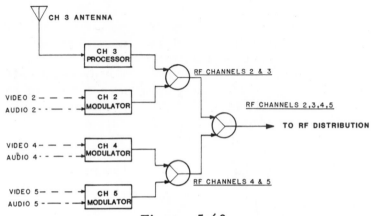

Figure 5.68
Combining Several RF Signals on One Cable

RF system design for more than about four destinations can be tricky. One usually starts by defining the desired signal level at the wall output connectors of the system. Where the TV set is connected, a level between 0 dBmV and +10 dBmV is usually expected (0 dBmV = 1 milliVolt of signal into a 75 Ohm load). Then the loss in each device (cable, tap, splitter, filter, etc.) is considered, as shown in Figure 5.69. To determine the necessary input power and amplification, the following formula can be used:

OUTPUT = INPUT - SYSTEM LOSSES

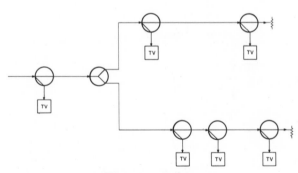

Figure 5.69
Typical RF Distribution System

Many RF signals may be combined and sent down one conductor pair by using different frequencies (different channels) selected from the table in Figure 5.70. The combining of the various channels must take place with equal signal powers to minimize interference among the various channels as the signals traverse the system.

CHANNEL	BANDPASS (MHz)	VISUAL CARRIER FREQUENCY (MHz)	AURAL CARRIER FREQUENCY (MHz)
2	54-60	55.25	59.75
3	60-66	61.25	65.75
4	66-72	67.25	71.75
5	76-82	77.25	81.75
6	82-88	83.25	87.75
7	174-180	175.25	179.75
8	180-186	181.25	185.75
9	186-192	187.25	191.75
10	192-198	193.25	197.75
11	198-204	199.25	203.75
12	204-210	205.25	209.75
13	210-216	211.25	215.75

Figure 5.70
North American TV Channel Allocations

The demodulator (tuner) in each TV receiver can select any of the channels for use. The benefits of being able to distribute several signals down a single cable with the capability of tuning one signal at the destination become obvious. There are drawbacks, however, because the technical limitations imposed on the video signal to meet RF channel bandwidth specifications make the picture resolution suffer. In addition, the processes of modulating and demodulating the baseband audio and video signals inherently introduce distortion into the output signals.

RF cables must match the balancing configurations and impedances of the signal to optimize the transfer of signal power. Coaxial cables with 75 Ohm characteristic impedance are used to carry unbalanced RF signals. Many television receivers are designed to use balanced RF signals from 300 Ohm "twin lead" cable. To match the balance configurations and impedances between the two systems, a balun (balanced/unbalanced transformer) similar to that shown in Figure 5.71 is used.

Figure 5.71
Typical TV Balun

ANCILLARY TELEVISION SIGNALS

Most of the time that the scan spends in the vertical interval is not visible in the picture. According to FCC Regulations, there is a maximum of 21 horizontal scans X 0.0000635 seconds/scan = 0.0013 seconds that is allowed no picture information, as shown in Figure 5.72. The instability of tube-type television equipment available when the specifications were adopted dictated such a wide tolerance when the rules were adopted; however, the stability of equipment has now achieved the point where the time spent in these scans can be reliably used for information that is not related to picture content.

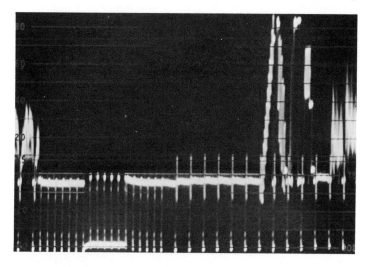

Figure 5.72
TV Signal Vertical Interval

VERTICAL INTERVAL TEST SIGNALS

Initially, the FCC required certain "vertical interval test signals," or "VITs" to be "inserted" in the vertical interval when operating a television transmitter by remote control. These VITs provide single horizontal scans of test signal information in the interval. To effectively use the VITs, waveform monitors are available to selectively display individual horizontal scans so that video signal distortions could accurately be measured. The measurements may be performed by using standard test signals while the normally viewable portions of the video signal remain undisturbed. VITs show the distortions that occur during the time of one horizontal scan and accurately reflects the distortions that may be occurring during any other horizontal scan in the picture. Distortions that have a duration equal to the time spent in a vertical scan cannot be measured.

As shown in Figure 5.73, VIT signals are generated as the program video is passed through a "VIT inserter." The VIT inserter strips any information on the selected lines and inserts the programmed test signals.

Figure 5.73
VIT System Block Diagram

VERTICAL INTERVAL REFERENCE SIGNALS

After the validity of VIT testing was confirmed in actual practice, the FCC approved the optional use of a special "Vertical Interval Reference Signal," or "VIR," to serve as a reference for correction of distortions that may occur to a video signal during transmission or recording. The VIR, inserted on Line 19 of both fields, suffers the same distortions as the remainder of the video signal. As shown in Figure 5.74, specific portions of the VIR are used to quantify errors in sync amplitude, luminance amplitude, chroma amplitude, pedestal amplitude, burst amplitude, and burst phase.

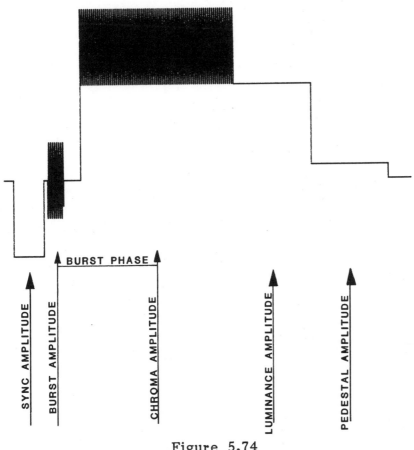

Figure 5.74
The VIR Signal

The VIR serves as a known, stable reference that establishes the quality of the video signal at the time of insertion. If the VIR is inserted into a good video signal, correction by VIR reference results in a good signal output. If the VIR is inserted into a bad video signal, correction by VIR reference results in a bad signal output.

As shown in Figure 5.75, optimum use of the VIR can be realized if the VIR is present on the camera output (which is assumed to produce a good video signal) and automatic correction takes place in the individual receiver or monitor. In this ideal system, all the referenced errors that may occur in recording, transmission, or reception of the video signal is automatically corrected by the receiver or monitor.

325

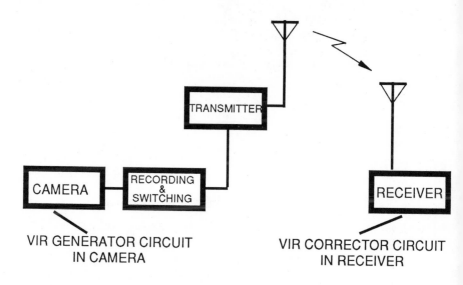

Figure 5.75
Optimum VIR System Block Diagram

OTHER VERTICAL INTERVAL SIGNALS

The latest development in the usage of the vertical interval allows insertion of information that is destined for audience use. The system looks similar to that shown in Figure 5.76. The information may be program related, as in "closed captioning" for the hearing impaired, "videotext" information, or "teletext" information that is totally unrelated to program content.

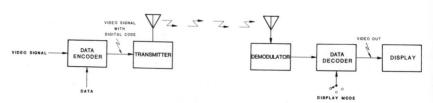

Figure 5.76
Block Diagram of a Typical System
for Inserting Ancillary Signals

In both systems, digital codes that represent the information unrelated to picture content are inserted at specific locations within the vertical interval. The digital codes are selectively detected and interpreted by a special "decoder" for display in the picture. Decoders allow the viewer the option of displaying the decoded information as an insert into the normal television picture or displaying full "pages" of text or graphics.

With the use of the vertical interval for "forward" communications, telephone lines or unviewable return CATV channels can be used for "reverse" communication. Such a system creates an interactive system that allows the viewer to request specific information from large, centrally-located computer data banks. This system also allows direct data communications from microcomputers to the large computer or to other microcomputers that may already be connected to the larger computer. This capability is often envisioned as providing an enormous computer resource in the hands of the general public. Forecasted uses include transportation information, telephone directory assistance, data "libraries," financial transactions, "shop-at-home" capabilities, and instructional services.

QUESTIONS

1. When editing component-format videotapes, how many transistor signal switches are activated when VTR A is selected? Must each signal be properly terminated when not in use?

2. In what two ways is a signal distorted as it travels through a coaxial cable?

3. What are the differences among a time base corrector (TBC), a field synchronizer, and a frame synchronizer?

4. What two visible effects will happen to a picture when the signal is not terminated? What two visible effects will happen to a picture when the signal has two terminations on the cable ("double terminated")?

5. What would be the effective termination value (in Ohms) of having ten loop-through picture monitors on the same signal path, each with an input impedance of 10,000 Ohms? What if the each of the picture monitors were terminated?

6. How many TV channels can be carried on an RF distribution system designed with a bandwidth of 360 MHz?

7. Does each cable of an RF distribution system need to be terminated?

APPENDICES

APPENDIX A

AUDIO PAD DESIGN PROCEDURE

The design of audio pads to provide a desired loss at given input and output impedances is a mathematically complex procedure. It is best performed with a calculator with scientific functions, and the procedure given here is structured accordingly.

STEP 1: Determine the amount of desired LOSS (in dB), impedances (Z_{in} and Z_{out}), and balancing configuration.

STEP 2: Calculate the minimum loss that a resistive pad can have when matching the two impedances given in STEP 1 by using the following formula:

$$\text{MINIMUM LOSS} = 20 \times \log \sqrt{\frac{Z_{max}}{Z_{min}}} - \sqrt{\frac{Z_{max}}{Z_{min}} - 1}$$

Where Z_{max} is the greater of Z_{in} or Z_{out} and Z_{min} is the lesser of Z_{in} or Z_{out}.

STEP 3: If the LOSS desired in STEP 1 is less than the MINIMUM LOSS of STEP 2, a pad with small enough loss cannot be designed or built. If the desired LOSS of STEP 1 is greater than the MINIMUM LOSS from STEP 2, proceed with preliminary calculations as follows:

333

$$k = 10^{\frac{LOSS}{20}}$$

$$A = \frac{k}{k^2-1}$$

$$B = \frac{k^2+1}{k^2-1}$$

$$C = 2 \times \sqrt{Z_{in} \times Z_{out}}$$

STEP 4: Using the factors calculated in STEP 3, calculate the resistor values for Figure A.1 by solving the following formulas:

$$R_3 = A \times C$$

$$R_1 = Z_{in} \times B - R_3$$

$$R_2 = Z_{out} \times B - R_3$$

If the audio system is balanced, the calculated resistance values are divided in half to be placed in the two halves of the balanced system.

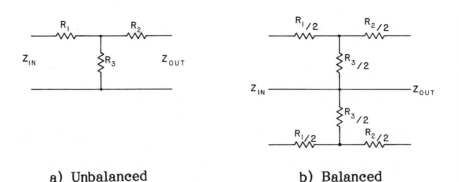

a) Unbalanced   b) Balanced

Figure A.1
Basic Schematics for Audio Pads

Appendix A

EXAMPLE 1:

   STEP 1: DESIRED LOSS = 10 dB

$$Z_{in} = 150 \text{ Ohms}$$

$$Z_{out} = 600 \text{ Ohms}$$

Unbalanced Configuration

   STEP 2:

$$\text{MINIMUM LOSS} = 20 \times \log \sqrt{\frac{600}{150}} - \sqrt{\frac{600}{150} - 1}$$

$$= -11.44 \text{ dB}$$

   STEP 3:

The pad cannot be designed because the desired LOSS is less than the MINIMUM LOSS.

EXAMPLE 2:

   STEP 1: DESIRED LOSS = 20 dB

$$Z_{in} = 150 \text{ Ohms}$$

$$Z_{out} = 600 \text{ Ohms}$$

Unbalanced Configuration

   STEP 2:

$$\text{MINIMUM LOSS} = 20 \times \log \sqrt{\frac{600}{150}} - \sqrt{\frac{600}{150} - 1}$$

$$= -11.44 \text{ dB}$$

STEP 3:

$$k = 10^{\frac{20}{20}} = 10^1 = 10$$

$$k^2 = 10^2 = 100$$

$$A = \frac{k}{k^2-1} = 0.10$$

$$B = \frac{k^2+1}{k^2-1} = 1.02$$

$$C = 2 \text{ X } \sqrt{Z_{in} \text{ X } Z_{out}} = 600$$

STEP 4: (See Figure A.2)

$$R_3 = A \text{ X } C = 600 \text{ X } 0.10 = 60 \text{ Ohms}$$

$$R_1 = Z_{in} \text{ X } B - R_3$$

$$= 150 \text{ X } 1.02 - 60 = 93 \text{ Ohms}$$

$$R_2 = Z_{out} \text{ X } B - R_3$$

$$= 600 \text{ X } 1.02 - 60 = 552 \text{ Ohms}$$

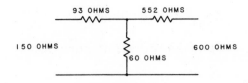

Figure A.2
Audio Pad Designed in Example 2

EXAMPLE 3:

   STEP 1: DESIRED LOSS = 5 dB

$$Z_{in} = 600 \text{ Ohms}$$

$$Z_{out} = 600 \text{ Ohms}$$

Balanced Configuration

   STEP 2:

$$\text{MINIMUM LOSS} = 20 \text{ X } \log \sqrt{\frac{600}{600}} - \sqrt{\frac{600}{600} - 1}$$
$$= 0 \text{ dB}$$

   STEP 3:

$$k = 10^{\frac{10}{20}} = 3.16$$

$$A = \frac{k}{k^2-1} = 0.32$$

$$B = \frac{k^2+1}{k^2-1} = 1.22$$

$$C = 2 \text{ X } \sqrt{600 \text{ X } 600} = 1200$$

   STEP 4: (See Figure A.3)

$$R_3 = 1200 \text{ X } 0.32 = 384 \text{ Ohms}$$

$$R_1 = 600 \text{ X } 1.02 - 384 = 228 \text{ Ohms}$$

$$R_2 = 600 \text{ X } 1.02 - 384 = 228 \text{ Ohms}$$

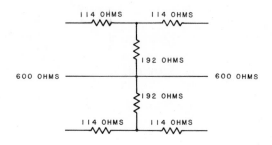

Figure A.3
Audio Pad Designed in Example 3

APPENDIX   B

8 mm

Signal Processing:

Heterodyne Chrominance (743 kHz)
Frequency Modulated Luminance (4.2 - 5.4 MHz)
Automatic Tracking

Tape

8mm Width
Metal Particle Formulation
0.565 inches-per-second across the transport (normal play)

Tracks

One helical-scan video track per field, alternating azimuth.
One helical-scan audio frequency modulation track.
Two *optional* longitudinal audio tracks.
Two *optional* pulse code modulation helical-scan audio tracks.

Typical Performance

Color Video/Picture:
230 TVL Horizontal Resolution
44 dB Signal -to- Noise Ratio

Longitudinal Audio/Sound:
Helical-scan AFM audio is standard equipment

Optional High-Fidelity Audio/Sound:
Pulse Code Modulation (optional)
20Hz-20kHz Bandwidth

8 mm

Optional Longitudinal Audio Track

Video Tracks

Optional PCM Audio Tracks

Optional Longitudinal Control Track

8mm

BetaMax®
(SMPTE Type G)

Signal Processing:
Heterodyne Chrominance (688 kHz)
Frequency Modulated Luminance
   (3.6-4.8 MHz, 4.4-5.6 MHz Super-Beta)
ED-Beta separates luminance and chrominance (Y/C) processing

Tape
1/2" Width
Oxide Formulation
2.93 inches-per-second across the transport (normal play)

Tracks
One helical-scan video track per field, alternating azimuth.
Two longitudinal audio tracks.
Two *optional* helical-scan audio tracks
   (Beta Hi-Fi, Super-Beta and ED-Beta).
One longitudinal control track.

Typical Performance
Color Video/Picture:
   240 TVL Horizontal Resolution
      (270 TVL Super-Beta, 500 TVL ED-Beta)
   45 dB Signal -to- Noise Ratio

Longitudinal Audio/Sound:
   50Hz-10kHz Bandwidth
   45dB Signal -to- Noise Ratio

Optional High-Fidelity Audio/Sound:
   Audio Frequency Modulation - AFM
      (Beta Hi-Fi, Super-Beta and ED-Beta)
   20Hz-20kHz Bandwidth

BetaMax®

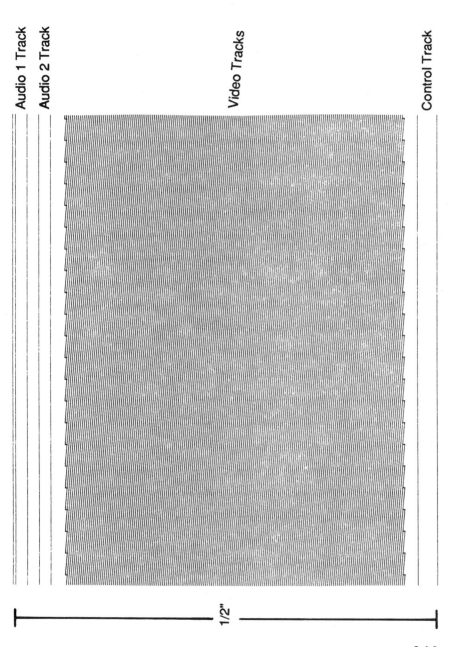

Audio 1 Track
Audio 2 Track
Video Tracks
Control Track

1/2"

VHS
(SMPTE Type H)

Signal Processing:
Heterodyne Chrominance (629 kHz Conventional, 629 kHz S-VHS)
Frequency Modulated Luminance (3.5-4.4 MHz Conventional, 5.4-7.0
S-VHS separates luminance and chrominance
(Y/C) processing throughout.

Tape
1/2" Width
Oxide Formulation
1.31 inches-per-second across the transport (normal play and S-VHS)

Tracks
One helical-scan video track per field, alternating azimuth.
Two longitudinal audio tracks.
Two *optional* helical-scan audio tracks (VHS Hi-Fi and S-VHS).
One longitudinal control track.

Typical Performance
Color Video/Picture:
240 TVL (400 TVL S-VHS) Horizontal Resolution
45 dB (47 dB S-VHS) Signal -to- Noise Ratio

Longitudinal Audio/Sound:
50Hz-12kHz Bandwidth
48dB Signal -to- Noise Ratio

Optional High-Fidelity Audio/Sound:
Depth Multiplexed (VHS Hi-Fi and S-VHS)
20Hz-20kHz Bandwidth

VHS

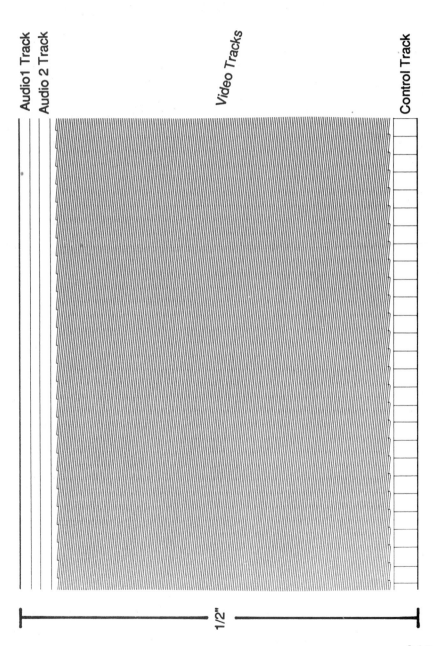

Audio1 Track

Audio 2 Track

Video Tracks

Control Track

1/2"

M-II

Signal Processing:

Time Compressed Component Chrominance (R-Y, B-Y)
Frequency Modulated Luminance (4.9 - 7.0 MHz)
Separate tape locations for Y, R-Y, and B-Y Components.

Tape

1/2" Width
Metal Particle Formulation
2.665 inches-per-second across the transport

Tracks

One helical-scan luminance video track per field.
One helical-scan chrominance (time compressed) video track per field.
Two longitudinal audio tracks.
Two *optional* audio frequency modulated helical-scan audio tracks.
One longitudinal address track.
One longitudinal control track.

Typical Performance

Color Video/Picture:

30 Hz-4.5 MHz (360 TVL) Luminance (Y) Horizontal Resolution
49 dB Luminance (Y) Signal -to- Noise Ratio

Longitudinal Audio/Sound:

50Hz-15kHz Bandwidth
56 dB Signal -to- Noise Ratio (Dolby®Off)

Optional High-Fidelity Audio/Sound:

Audio Frequency Modulation - AFM
Pulse Code Modulation
20Hz-20kHz Bandwidth

M-II

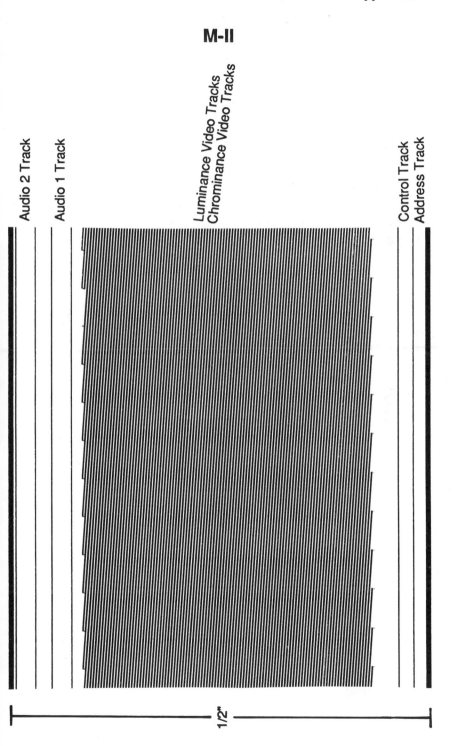

Audio 2 Track

Audio 1 Track

Luminance Video Tracks
Chrominance Video Tracks

Control Track
Address Track

1/2"

BetaCam®

Signal Processing:
Time Compressed Component Chrominance (R-Y, B-Y)
Frequency Modulated Luminance (4.4 - 6.4 MHz, 5.7 - 7.7 MHz SP)
Separate tape locations for Y, R-Y, and B-Y Components.

Tape
1/2" Width
Oxide Formulation
4.677 inches-per-second across the transport

Tracks
One helical-scan luminance video track per field.
One helical-scan chrominance (time compressed) video track per field.
Two longitudinal audio tracks.
Two *optional* pulse code modulated helical-scan audio tracks.
One longitudinal address track.
One longitudinal control track.

Typical Performance
Color Video/Picture:
30 Hz-4.5 MHz (360 TVL) Luminance (Y) Horizontal Resolution
51 dB Luminance (Y) Signal -to- Noise Ratio

Longitudinal Audio/Sound:
50Hz-15kHz Bandwidth
54 dB Signal -to- Noise Ratio (Dolby®Off)

Optional High-Fidelity Audio/Sound:
Audio Frequency Modulation - AFM
Pulse Code Modulation
20Hz-20kHz Bandwidth

BetaCam®

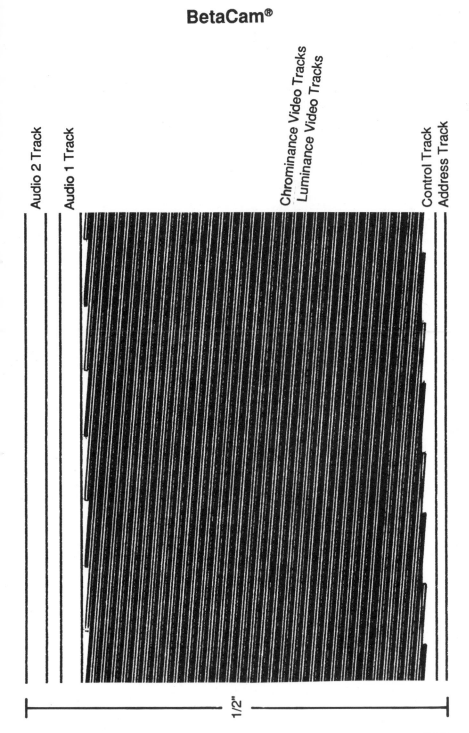

Audio 2 Track

Audio 1 Track

Chrominance Video Tracks

Luminance Video Tracks

Control Track

Address Track

1/2"

3/4" U-Matic®
(SMPTE Type E)

Signal Processing:

Heterodyne Chrominance (688 kHz Conventional, 688 kHz SP)
Frequency Modulated Luminance

(3.8-5.4 MHz Conventional, 5.0-6.6 MHz SP)

Tape

3/4" Width
Oxide Formulation
3-3/4 inches-per-second across the transport

Tracks

One helical-scan video track per field.
Two longitudinal audio tracks.
One *optional* longitudinal address track.
One longitudinal control track.

Typical Performance

Color Video/Picture:
260 TVL (340 TVL SP) Horizontal Resolution
47 dB (47 dB SP) Signal -to- Noise Ratio

Audio/Sound:
50Hz-15kHz ±4dB (±3dB SP) Bandwidth
<2% Distortion
50dB (52dB SP) Signal -to- Noise Ratio
(3% Distortion, no noise reduction)

3/4" U-Matic®

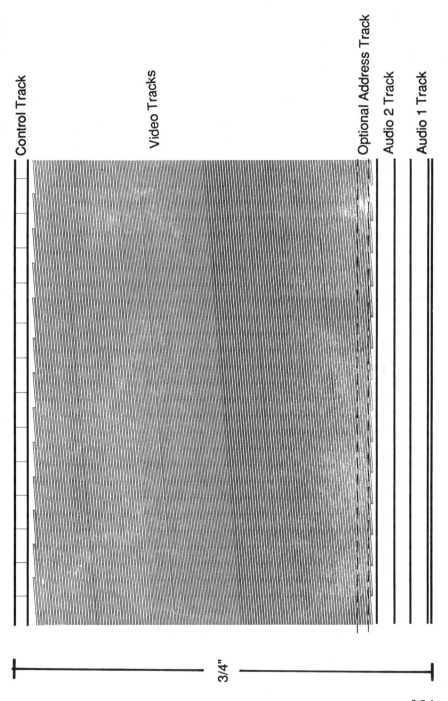

Control Track

Video Tracks

Optional Address Track

Audio 2 Track

Audio 1 Track

3/4"

D-1

Signal Processing:

Digitized Component Chrominance (R-Y, B-Y)
   6.75 million samples-per-second, R-Y and B-Y
Digitized Luminance
   13.5 million samples-per-second, Y
Separate tape locations for Y, R-Y, and B-Y Components.

Tape

19mm Width
Metal Particle Formulation
11.283 inches-per-second across the transport

Tracks

One helical-scan luminance video track per field.
One longitudinal audio track.
Four helical-scan audio tracks.
One longitudinal address track.
One longitudinal control track.

Typical Performance

Color Video/Picture:

30 Hz-5.75 MHz (460 TVL) Luminance (Y) Horizontal Resolution
56 dB Luminance (Y) Signal -to- Noise Ratio

Longitudinal Audio/Sound:

100Hz-10kHz Bandwidth
50 dB Signal -to- Noise Ratio (Dolby®Off)

Optional High-Fidelity Audio/Sound:

Pulse Code Modulation
20Hz-20kHz Bandwidth

D-1

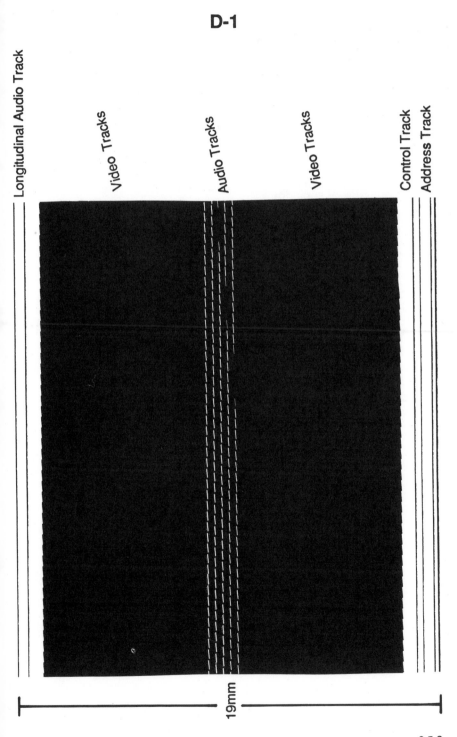

Longitudinal Audio Track

Video Tracks

Audio Tracks

Video Tracks

Control Track

Address Track

19mm

D-2

Signal Processing:
Digitized   Composite Video
  14,318,000 samples-per-second

Tape
19mm Width
Metal Particle Formulation
5.185 inches-per-second across the transport

Tracks
One helical-scan luminance video track per field.
One longitudinal audio track.
Four helical-scan digitized audio tracks.
One longitudinal address track.
One longitudinal control track.

Typical Performance

Color Video/Picture:
30 Hz-5.75 MHz (440 TVL) Horizontal Resolutio
54 dB Luminance Signal -to- Noise Ratio

Longitudinal Audio/Sound:
NOTE: DIGITIZED AUDIO IN HELICAL TRACKS IS STANDARD
100Hz-12kHz Bandwidth
44 dB Signal -to- Noise Ratio (Dolby®Off)

Optional High-Fidelity Audio/Sound:
Digitized Audio (Standard Capability)
20Hz-20kHz Bandwidth

D-2

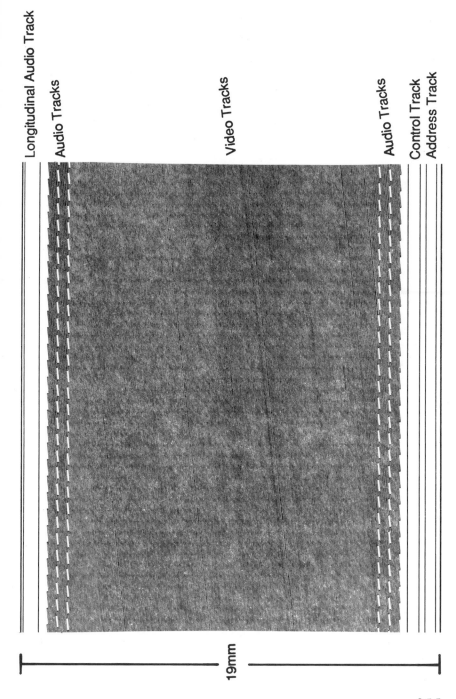

Longitudinal Audio Track

Audio Tracks

Video Tracks

Audio Tracks

Control Track

Address Track

19mm

1" - C

Signal Processing:
Direct Frequency Modulation of Composite Signal (7.06 - 10.00 MHz)
Automatic Tracking

Tape
1" Width
Oxide Formulation
9.6 inches-per-second across the transport

Tracks
One helical-scan video track per field.
Three longitudinal audio tracks.
Two *optional* pulse code modulation helical-scan audio tracks.

Typical Performance
Color Video/Picture:
30 Hz-5.0 MHz (400 TVL) Luminance (Y) Horizontal Resolution
46 dB Luminance (Y) Signal -to- Noise Ratio

Longitudinal Audio/Sound:
50Hz-15kHz Bandwidth
56 dB Signal -to- Noise Ratio

Optional High-Fidelity Audio/Sound:
Audio Frequency Modulation - AFM
Pulse Code Modulation

1" - C

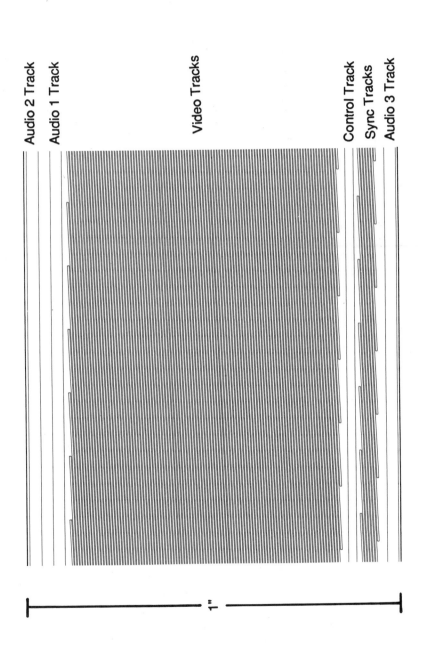

Audio 2 Track
Audio 1 Track
Video Tracks
Control Track
Sync Tracks
Audio 3 Track

1"

2" "Quad"

Signal Processing:
Direct Frequency Modulation of Composite Signal
Automatic Tracking

Tape
2" Width
Oxide Formulation
15 inches-per-second across the transport

Tracks
One transverse-scan video track per 15 horizontal scans.
Two longitudinal audio tracks.
One longitudinal address track.

Typical Performance

Color Video/Picture:
30 Hz-5.0 MHz (400 TVL) Luminance (Y) Horizontal Resolution
46 dB Luminance (Y) Signal -to- Noise Ratio

Longitudinal Audio/Sound:
50Hz-15kHz Bandwidth
56 dB Signal -to- Noise Ratio

Optional High-Fidelity Audio/Sound:
Pulse Code Modulation

2" "Quad"

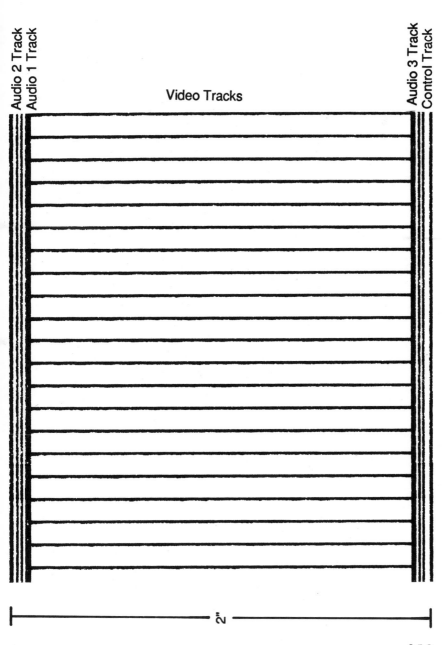

Audio 2 Track
Audio 1 Track

Video Tracks

Audio 3 Track
Control Track

2"

APPENDIX C

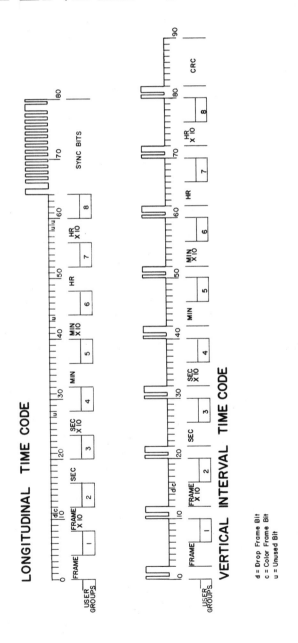

LONGITUDINAL TIME CODE

VERTICAL INTERVAL TIME CODE

d = Drop Frame Bit
c = Color Frame Bit
u = Unused Bit

APPENDIX D

Basic system tests are often necessary to ensure optimum facility performance and maintenance. The basic tests outlined here can measure problems that directly affect the quality of the finished product and may be considered technical areas where minimal operating specifications should be developed. There are many, many more detailed procedures and tests, but most are beyond the equipment or capabilities of the average corporate television facility. The specifications given in these test procedures do not correspond to any particular industry standard, but are intended to indicate an acceptable performance range.

I. VIDEO TESTS

   A. VIDEO SIGNAL LEVEL

      1. Specification: The voltage of a video signal shall optimally be 1 Volt peak-to-peak into a 75-Ohm load. In North America, the signal shall be comprised of 0.714 Volts of picture information and 0.286 Volts of sync information. The color burst shall optimally be 0.286 Volts peak-to-peak.

      2. Equipment Required (connected as shown in Figure D.1):

         a. Video signal generator with a calibrated 1 Volt peak-to-peak output signal

         b. Waveform monitor (calibrated)

         c. Precision 75-Ohm termination

3. Measurement: From sync tip to peak white should be 140 IRE (1 Volt). From sync tip to blanking level should be 40 IRE (0.286 Volts). From blanking level to peak white level should be 100 IRE (0.714 Volts). The burst should be 40 IRE from top to bottom and centered vertically on the blanking level.

4. Visible Results: If the levels are incorrect, too much or too little picture contrast will be evident. In extreme cases, a video monitor will lose horizontal and/or vertical lock, particularly with dark pictures.

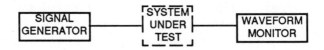

Figure D.1
Equipment Connection to Measure
Video Signal Level

B. DIFFERENTIAL VIDEO GAIN

1. Specification: The differential gain of system shall not exceed 15%.

2. Equipment Required (connected as shown in Figure D.2):

   a. Video signal generator with a modulated stairstep output signal.

   b. Vectorscope or waveform monitor (2H Display)

   c. Precision 75-Ohm termination

3. Measurement: The peak-to-peak voltage in the modulation levels are examined for changes. Differential gain is expressed as the maximum difference of the signal. Many waveform monitors have a "DIFFERENTIAL GAIN" or "CHROMA" mode that displays only the modulation information, stripping away all the rest of the signal.

Differential gain can also be measured on a vectorscope by adjusting the GAIN of the vectorscope so the dot generated by the modulation (not the burst dot) lies on the graticule circle. The amount of differential gain can then be measured by examining and measuring the dot for any elongation of the dot along the axis from the center of the circle.

Some vectorscopes use sophisticated techniques with more accurate results, but the principle of measuring changes in modulation levels still apply.

4. Visible Results: If there is appreciable differential gain distortion, the color in the picture will lose or gain saturation as the brightness is increased or decreased.

a) With Waveform Monitor

b) With Vectorscope

Figure D.2
Equipment Connection to Measure
Differential Gain

365

C. DIFFERENTIAL VIDEO PHASE

1. Specification: The differential phase of a system shall not exceed 5°.

2. Equipment Required (connected as was shown in Figure D.2.b.):

    a. Video signal generator with modulated stairstep output signal. (The modulation is of constant phase on the generated signal.)

    b. Vectorscope

    c. Precision 75-Ohm termination

3. Measurement: The GAIN of the vectorscope is adjusted so the dot generated by the modulation (not the Burst dot) lies on the graticule circle. The amount of differential phase can then be measured as the number of degrees the dot is distorted from a perfect circle around the circular axis.

   Some vectorscopes use sophisticated techniques with more accurate measurement results, but the principle of measuring change in modulation phase remains constant.

4. Visible Results: If there is excessive differential phase, there will be a change of hue as the picture brightness is varied.

D. VIDEO FREQUENCY RESPONSE

1. Specification: The frequency response of a television system shall be +0.5,-0.9 dB; 100 Hz to 4.2 MHz.

2. Equipment Required (connected as was shown in Figure D.1):

a. Video signal generator with a multiburst output signal

b. Waveform monitor (2H Display)

c. Precision 75-Ohm termination

3. Measurement Technique: Adjust the waveform monitor GAIN for 1 Volt peak-to-peak display from sync tips to peak white (140 IRE in North America). Measure the maximum difference in peak-to-peak levels of the Burst packets. The difference is expressed in dB and may be calculated by using the formula:

dB Response =

$$20 \text{ X LOG} \frac{\text{(Max p-p Level)}-\text{(Min p-p Level)}}{\text{(Max p-p Level)}}$$

4. Visible Results: In general, the better the response at high frequencies, the more detail is evident in the picture (corresponding to resolution); however, if the response is too great at the higher frequencies, sharp black-to-white transitions will tend to oscillate and produce ghosts. If response is poor at the low frequencies, the picture will tend to "smear" together.

E. VIDEO HUM

1. Specification: The hum (or "tilt") in a television system shall not exceed 4.0%.

2. Equipment Required (connected as was shown in Figure D.1):

a. Video signal generator with a 1 Volt peak-to-peak output signal

b. Waveform monitor (2V Display)

c. Precision 75-Ohm termination

3. Measurement Technique: Measure the peak-to-peak baseline shift of the signal as a percentage of the total signal (1 Volt).

4. Visible Effects: If excessive hum is present, a horizontal bar of reduced picture brightness will roll through the display.

II. AUDIO

A. AUDIO SIGNAL LEVEL

1. Specification: The audio level at any given point shall be determined by the optimum input level of a piece of equipment. Most systems use 0 dBm or +4 dBm as a standard.

2. Equipment Required (connected as shown in Figure D.3):

   a. Audio tone generator with appropriate level at 400 Hz

   b. AC voltmeter with high input impedance and high sensitivity

   c. Appropriately sized termination

Figure D.3
Equipment Connection to Measure
Audio Signal Level

3. Test Procedure: Apply a signal from the tone generator as the designed level to the input of the path under test. Measure the output of the path under test with the AC voltmeter.

4. Audible Effects: If the level of the signal at each point in the path is drastically different from the design level, there may not be enough apparent volume of the output signal (with a low level) or too much volume and distortion (with a high level).

B. AUDIO SIGNAL-TO-NOISE

1. Specification: The signal-to-noise ratio shall be greater than 55 dB, 50 Hz - 15 KHz.

2. Equipment Required (connected as shown in Figure D.4):

   a. Tone generator

   b. AC voltmeter

   c. Appropriately sized termination

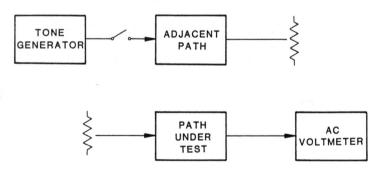

Figure D.4
Equipment Connection to Measure
Audio Signal-to-Noise Ratio

3. Test Procedure: Adjust the system for proper levels throughout. Measure the output signal level in dBm. Replace the tone generator with the terminating resistor. Measure the amount of noise generated in the system with the input terminated. The signal-to-noise ratio (in dB) equals the measured signal level (in dBm) minus the measured noise (in dBm).

4. Audible Effects: If the signal-to-noise ratio is too low, there will be an objectionable constant "hiss" in the output signal.

C. AUDIO FREQUENCY RESPONSE

1. Specification: The audio frequency response shall vary less than ± 1.0 dB from 60Hz to 12kHz.

2. Equipment Required (connected as shown in Figure D.5):

   a. Variable-frequency tone generator

   b. AC Voltmeter

   c. Appropriately sized termination

3. Test Procedure: Measure and record the audio signal level at the input and at the output of the system under test at the following frequencies:

   60 Hz
   400 Hz
   1 kHz
   5 kHz
   10 kHz
   12 kHz

   For each of the measurement frequencies, subtract the input level from the output level to determine the gain of the system under test. This gain figure (in dB) should not vary by more than the maximum variation specified.

4. Audible Effects: If the frequency response varies excessively, certain frequencies (pitches) in sound will be louder (or softer) than the rest of the signal. An analogy is the way that a voice sounds through a telephone circuit where the audio frequency response is deliberately altered to conserve bandwidth.

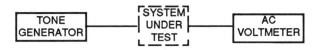

Figure D.5
Equipment Connection to Measure
Audio Frequency Response

D. AUDIO DISTORTION

1. Specification: Audio distortion shall be less than 2% at 1kHz and rated signal level.

2. Equipment Required:

    a. Low-distortion tone generator (may be within distortion measurement test set)

    b. AC voltmeter (may be within distortion measurement test set)

    c. Distortion meter (may be within distortion measurement test set)

    d. Appropriately sized terminations

3. Test Procedure: Adjust the tone generator for the specified output frequency and the level specified as the rated input of the system under test. Adjust the output of the system under test to the level specified as the rated output of the system under test. Measure and record the distortion (in %) displayed on the distortion meter.

Figure D.6
Equipment Connection to Measure Audio Distortion

APPENDIX   E

COMMON INTERNATIONAL TELEVISION STANDARDS

CCIR STANDARDS	SCANNING		CHROMINANCE		
	HORIZ. SCANS -PER- FRAME	FRAMES -PER- SECOND	NTSC Amplitude and Phase Modulated	PAL Amplitude and Phase Modulated	SECAM Frequency Modulated
B,G,D,H,I,L,K	625	50	-	4.43361875 MHz	4.40625 & 4.4250MHz
M	525	59.94	3.57954545 MHz	3.57561149 MHz	-

NOTES:

1. It is common practice to produce in NTSC-M and transcode to comply with the PAL-M transmission standard.

2. It is common practice to produce in PAL-B,G and transcode to comply with the SECAM transmission standards.

3. Differences between B,G,D,H,I,K, and L transmission standards do not affect television production standards.

4. Derivation of luminance (Y) is the same for all standards.

5. For more information, consult CCIR Report 524-2.

TELEVISION TRANSMISSION STANDARDS
FOR SELECTED COUNTRIES

COUNTRY	STANDARD(S)
Australia	PAL-B
Austria	PAL-B
Bahrain	PAL-B
Belgium	PAL-B,H
Brazil	PAL-M
Canada	NTSC-M
Chile	NTSC-M
China	PAL-D
Columbia	NTSC-M
Egypt	SECAM-B
France	SECAM-L
Germany (East)	SECAM-B,G
Germany (West)	PAL-B,G
Hong Kong	PAL-B,I

COUNTRY	STANDARD(S)
Japan	NTSC-M
Korea (South)	NTSC-M
Mexico	NTSC-M
New Zealand	PAL-B
Peru	NTSC-M
Saudi Arabia	SECAM-B,G
Singapore	PAL-B
South Africa	PAL-I
Switzerland	PAL-B,G
Taiwan	NTSC-M
United Kingdom	PAL-I
United States	NTSC-M
USSR	SECAM-D,K
Venezuela	NTSC-M

RS-189A COLOR BARS
("Split-Field" Color Bars)

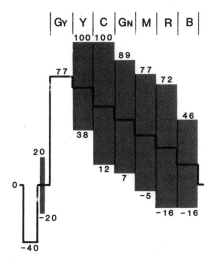

Top
of
Picture

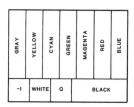

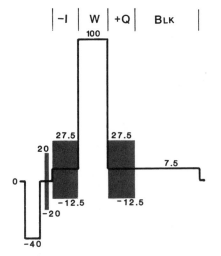

Bottom
of
Picture

FCC Part 73 Excerpt
(Television Broadcast Specifications)

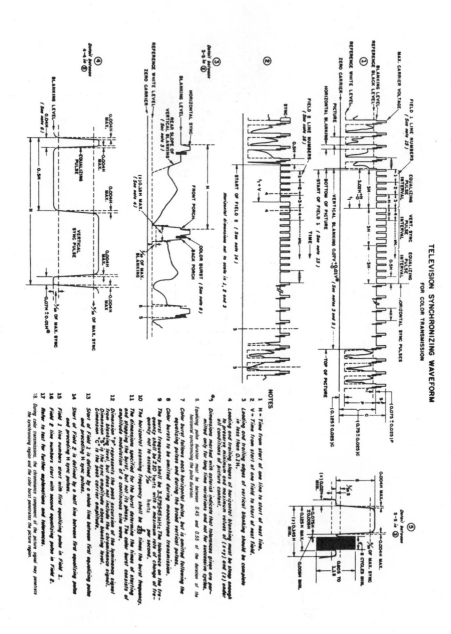

BIBLIOGRAPHY

Bartlet, George W. (ed.). *National Association of Broadcasters Engineering Handbook*. 6th ed. Washington, D.C.: National Association of Broadcasters, 1975.

Bensinger, Charles. *The Video Guide*. 2nd ed. rev. Santa Fe, New Mexico: Video-Info Publications, 1981.

Beynon, J.D.E. and Lamb, D.R. (ed.). *Charge-coupled devices and their applications*. St. Louis: McGraw-Hill Book Company (UK) Limited, 1980.

Benson, K. Blair (ed. in chief). *Television Engineering Handbook*. New York: McGraw-Hill Book Company, Inc., 1986.

Cartwright, Steve R. *Training with Video*. White Plains, New York: Knowledge Industry Publications, Inc., 1986.

Castleman, Kenneth R. *Digital Image Processing*. Englewood Cliffs, New Jersey: Prentice-Hall, Inc., 1979.

Crutchfield, E.B. (ed.). *National Association of Broadcasters Engineering Handbook*. 7th ed. Washington, D.C.: National Association of Broadcasters, 1985.

Cunningham, John E. *Cable Television*. 2d ed. Indianapolis: Howard W. Sams & Co., Inc., 1980.

Ennes, Harold E. *Digitals in Broadcasting*. Indianapolis: Howard W. Sams & Co., Inc., 1977.

Ennes, Harold E. *Television Broadcasting: Equipment, Systems, Operating Fundamentals*. 2d ed. Indianapolis: Howard W. Sams & Co., Inc., 1979.

Ennes, Harold E. *Television Broadcasting: Tape Recording Systems*. 2n ed. Indianapolis: Howard W. Sams & Co., Inc., 1979.

Ennes, Harold E. *Television Broadcasting: Systems Maintenance.* 2n ed. Indianapolis: Howard W. Sams & Co., Inc., 1978.

Fredida, Sam and Malik, Rex. *The Viewdata Revolution.* London: Associated Business Press, 1979.

GTE Sylvania. *Lighting Handbook for Television, Theatre, Professional Photography.* 6th ed. Danvers, Massachusetts: GTE Sylvania Incorporated; Lighting Center, 1978.

Giloi, Wolfgang K. *Interactive Computer Graphics: Data Structures, Algorithms, Languages.* Englewood Cliffs, New Jersey: Prentice-Hall, Inc., 1978.

Hansen, Gerald L. *Introduction to Solid-State Television Systems: Color and Black & White.* Englewood Cliffs, New Jersey: Prentice-Hall, Inc., 1969.

Harris, Cyril M. (ed). *Handbook of Noise Control.* (2d ed.). New York: McGraw-Hill Book Company, 1979.

Hutson, Geoffrey H. *Colour Television Theory: PAL-System Principles and Receiver Circuitry.* St. Louis: McGraw-Hill Publishing Company (UK) Limited, 1971.

Kiver, Milton S. and Kaufman, Milton. *Television Electronics: Theory and Servicing.* 8th ed. New York: Van Nostrand Reinhold Company, Inc., 1983.

Kybett, Harry. *Video Tape Recorders.* 2n ed. Indianapolis Howard W. Sams & Co., Inc., 1978.

Lancaster, Don. TV *Typewriter Cookbook.* Indianapolis: Howard W. Sams & Co., Inc., 1976.

LeTourneau, Tom. *Lighting Techniques for Video Production: The Art of Casting Shadows.* White Plains, New York: Knowledge Industry Publications, Inc., 1987.

Marsh, Ken. *Independent Video: A Complete Guide to the Physics, Operation, and Application of the New Television for the Student, the Artist, and for Community TV.* New York: Simon and Schuster, 1974.

Martin, James. *Telematic Society: A Challenge for Tomorrow.* Englewood Cliffs, New Jersey: Prentice-Hall, Inc., 1981.

Mathias, Harry and Patterson, Richard. *Electronic Cinematography.* Belmont, California: Wadsworth Publishing Company, Inc., 1985.

McGinty, Gerald P. *Video Cameras: Operation and Servicing.* Indianapolis: Howard W. Sams & Company, Inc., 1984.

378

McQuillin, Lon. *The Video Production Guide*. Indianopolis: Howard W. Sams & Co., Inc.,1983.

National Fire Protection Association. *National Electrical Code 1984 Edition*. Quincy, Massachusetts: National Fire Protection Association, 1983.

Paulson, C. Robert. (Principal Author). *BM/E's ENG/EFP/EPP Handbook: Guide to Using Mini Video Equipment*. New York: Broadband Information Services, Inc., 1981.

Remley, Frederick M. (ed.). *One-Inch Helical Video Recording*. Scarsdale, N.Y.: Society of Motion Picture and Television Engineers, 1978.

Robinson, Richard. *The Video Primer: Equipment, Production, and Concepts*. New York: Quick Fox Books, 1974.

Sigel, Efrem, Schubin, Mark, and Merrill, Paul F. *Video Discs: The Technology, the Applications and the Future*. New York: Van Nostrand Reinhold Company, 1980.

Sigel, Efrem (ed.). *Videotext: The Coming Revolution in Home/Office Information Retrieval*. New York: Harmony Books, 1980.

Tremaine, Howard M. *Audio Cyclopedia*. 2n ed. Indianapolis: Howard W. Sams & Co., Inc., 1973.

Tremaine, Howard M. *Passive Audio Network Design*. Indianapolis: Howard W. Sams & Co., Inc., 1964.

Utz, Peter. *Video User's Handbook*. Englewood Cliffs, New Jersey: Prentice-Hall, Inc., 1980.

Zetl, Herbert. *Television Production Handbook*. 4th ed. Belmont, California: Wadsworth Publishing Company, Inc., 1984.

INDEX

Vertical drive, 202
Vertical interval
  Reference (VIR), 324-26
  switching, 180, 187,263-64
  sync pulse, 149
  Test (VIT), 322-324
  Time Code (VITC), 187
Vertical sync pulse, 149
VHS, 146-47
Video discs, 194-95
Video signal, 7, 21ff, 29-30
Video switcher, 261
  auxiliary inputs, 267
  buses, 265-67
  cut bar, 266
  key effects, 272ff
  reentry, 276-78
  special effects, 269ff
  transitions, 268ff
  vertical interval switching,
    263-65
Video tape, 190-93
  cleaning cassettes, 192
  damage, 156-57, 193
  drop outs, 190-91
  duplication, 171-74
  editing, 174ff
  lubricant, 193
Video tape duplication, 171-74
Video tape formats, 146
  BetaCam®, 167, 348-49
  Betamax®, 164, 342-44
  Betamax Hi-Fi®, 171,342-44
  D-1, 168-170, 352-53
  D-2, 168-70, 354-55
  ED Beta®, 165, 342-44
  8mm, 157, 164, 340-41
  M-II, 32, 167, 346-47
  1"-C, 157, 159-160, 356-57
  3/4" 157, 164, 350-51
  3/4" SP, 164, 350-51
  2",157,32, 159-160, 358-59
  Super VHS, 165, 344-45
  VHS, 146-47, 164, 344-45
  VHS Hi-Fi, 170-71, 344-45
Video Tape Recorder (VTR)
  audio dub, 174
  audio tracks, 146

Video Tape Recorder *(cont.)*
  cleaning, 192
  capstan servo, 142, 150-52
  color dub, 174
  component recording,
    167-70
  confidence head, 142
  control track, 150, 182-83,
    186
  dew sensor, 193
  digital recording, 168-70
  drum servo, 149-50
  dub mode, 173-174
  dynamic tracking, 154
  editing, 174ff
  editing recorders, 179ff
  end-of-tape circuit, 158
  erase mode, 140
  FM dub, 174
  frame servo, 180-81
  head clog, 191-92
  head switch, 150
  helical scan, 145
  heterodyne recording,
    160ff
  interchange, 146
  magnetic heads, 140-42
  pause mode, 158
  pinch roller, 142
  play mode, 140
  record mode, 137-140
  reel motors, 142
  servo circuits, 148ff
  skew control, 154-56
  tension, 142-43, 154-56
  time code, 186-89
  tracking, 152-54
  video heads, 144
  writing speed, 145
  Y/C, 165-66
  Y/688, 173
Videotext, 326
Vidicon, 35-36
VIR (Vertical Interval Reference),
  324-26
VIT (Vertical Interval Test), 322-324
VITC (Vertical Interval Time Code),187
Volume Unit (VU), 122